THE DYNAMICS OF
Arthropod
Predator-Prey
Systems

MONOGRAPHS IN POPULATION BIOLOGY

EDITED BY ROBERT M. MAY

THE DYNAMICS OF
Arthropod Predator-Prey Systems

MICHAEL P. HASSELL

PRINCETON, NEW JERSEY
PRINCETON UNIVERSITY PRESS

1978

Copyright © 1978 by Princeton University Press
Published by Princeton University Press, Princeton, New Jersey
In the United Kingdom: Princeton University Press, Guildford, Surrey

ALL RIGHTS RESERVED

Library of Congress Cataloging in Publication Data will be
found on the last printed page of this book
This book has been composed in Linofilm Baskerville
Clothbound editions of Princeton University Press books
are printed on acid-free paper, and binding materials are
chosen for strength and durability.
Printed in the United States of America by Princeton
University Press, Princeton, New Jersey

Preface

Predator-prey interactions were first modeled more than half a century ago, but only recently have these models become markedly more sophisticated. This book surveys part of this contemporary scene, considering a particular kind of interaction—that of arthropod predators and their prey (or parasitoids and their hosts). In choosing arthropods as our subjects, the difference equation models that are a recurrent theme throughout the book become especially appropriate, since so many arthropods, at least within temperate regions, tend to have fairly discrete generations.

The important components of predator-prey models that are considered in some detail include (1) a density dependent, rather than constant, prey rate of increase, (2) various forms of functional response, (3) non-random search and predator interference, and (4) a predator rate of increase that is a function of the factors affecting predator survival, developmental rate, and fecundity. We then turn to more complex systems, involving in the first place polyphagous predators and how they can affect the coexistence of competing prey species, and secondly, interactions with more than one predator species, where the emphasis is on the conditions enabling the predators to coexist. The final chapter is devoted to biological control and in particular to the ways in which models discussed earlier in the book can contribute to a theoretical basis for biological control practices.

The emphasis throughout is on the correspondence between models and data, and on the dynamical effects of each component of predation considered. The data have been culled more or less directly from several sources, which accounts for the seemingly capricious use of error bars; where they were used in the original, they have been reproduced here. Mathematical details have to a large extent been omitted from the main text, and the more mathematically inclined reader is referred to one of the Appendices or to the appropriate original papers.

I am deeply indebted to many people for their help in the preparation of this volume and during its gestation. Foremost among these is Robert May. His inspiration permeates much of the book, and I am particularly grateful to him for giving me free rein to use his model for predator aggregation in Chapters 4 and 5 and for his considerable help with Chapter 8 and the Appendices.

Other colleagues whose ideas have helped shape my own are R. Anderson, J. R. Beddington, T. Bellows, M. J. Crawley, G. R. Conway, H. N. Comins, M. J. W. Cock, J. R. Krebs, J. H. Lawton, S. McNeill, V. C. Moran, D. J. Rogers, T. R. E. Southwood, T. Palmer, G. C. Varley and J. Waage. Amongst these, J. H. Lawton and T. R. E. Southwood were of especial help in reviewing the manuscript and offering plentiful and excellent advice.

Finally, my gratitude is also due to Carole Collins for typing the manuscript, and to my wife for her encouragement and help in preparing the figures.

Imperial College, 1977   M.P.H.

Contents

THE DYNAMICS OF
Arthropod
Predator-Prey
Systems

CHAPTER ONE

Introduction

Theoretical population ecology emerged as a well-founded discipline between 1920 and 1935. Pearl and Reed (1920) rediscovered Verhulst's (1838) "logistic" model for single-species population growth, Lotka (1925) and Volterra (1926) developed a model for two-species competition, and Thompson (1924), Lotka (1925), Volterra (1926), Nicholson (1933), and Nicholson and Bailey (1935) produced models for predator-prey interactions. Only mutualism was neglected as an important species interaction affecting a population's dynamics.

The principal aim in studying the dynamics of these interactions is to explain the distribution and abundance of animal populations; a vast undertaking, encompassing most aspects of population ecology. Unfortunately, predator-prey and competitive interactions under field conditions cannot be easily categorized. Thus, some predators have little discernible effect on their prey's dynamics (e.g. Hassell, 1969); in other interactions the predators are clearly maintaining their prey at very low equilibrium levels (e.g. Huffaker and Kennett, 1966; DeBach, Rosen, and Kennett, 1971); a few predators and their prey undergo cyclic oscillations (e.g. Lack, 1954; Miller, 1966); and some interactions are characterized by episodic prey outbreaks when predation ceases to be limiting (e.g. Schwertfeger, 1935; Clark, 1963). Within the confines of a laboratory experiment, however, many of the vagaries of the real world are eliminated and the outcomes of interactions are more simply classified. For instance, three types of predator-prey interaction have been observed, differing in their stability properties: (1) those where the prey becomes extinct (e.g. Gause, 1934; Luckinbill, 1973); (2) those where both populations oscillate out of phase with each other (e.g. Huffaker, 1958; Huffaker, Shea, and Herman, 1963); and (3) those where both populations persist, but fluctuate less regularly (e.g. Utida, 1957; Burnett, 1977). Theoretical predator-prey studies would be well served by focusing in the first place on these simple single predator-single prey systems and especially on the factors affecting the stability and equilibrium levels of the interacting populations. Models are needed in which the essential components of predator search and reproduction are included, and in which the dynamic effects of each component are known. Such models will then provide the basis for understanding the "noisy" and inevitably more complex field situations and for addressing such broader problems as the role of predators within a community of potential prey species.

Despite the interest aroused by the early classical studies, little attempt was made in the ensuing years to collect the evidence necessary to confirm, reject, or modify the assumptions upon which they rest. Certainly, Gause (1934) performed valuable experiments on two-species interactions, but these can only be used to test the gross predictions of the models rather than the functional relationships upon which the models are based. Until recently (see Gilpin and Justice, 1972 and Gilpin and Ayala, 1973), Crombie's (1944, 1945, 1946) experiments with stored product beetles and moths stood out as the only critical attempt to test some of the assumptions of the Lotka-Volterra competition model. Predator-prey models were similarly neglected. Although DeBach and Smith (1941a, b; 1947), Ullyett (1949a, b), and Burnett (1951, 1954, 1956, 1958a) performed useful laboratory experiments, these were only casually related to existing models, and Varley's (1947) pioneering application of the Nicholson-Bailey model to field data was not concerned with testing the model's assumptions.

The real turning point for theoretical studies on predation came with the classic works of Holling (1959a, b), Watt (1959), and Ivlev (1961). They argued against the basic tenet of the Lotka-Volterra and Nicholson-Bailey models —that the attack rate per predator is a linear function of prey density—and presented new relationships in which attack rates rise monotonically towards a maximum as prey density increases. Watt, in addition, permitted predator searching efficiency to decline with increases in the density of searching predators. The distinguishing feature of these studies is that the generality of the new sub-models was confirmed from several sets of experimental data, Ivlev working with fish and Holling and Watt drawing largely on the previous insect studies of DeBach and Smith, Ullyett, and Burnett. So began a period in which experimental studies on predation became commonplace and sufficient to allow the development of more detailed, yet general predator-prey models. Future developments are unlikely to come from predator-prey models that are completely divorced from this body of empirical information.

Much of this experimental activity has been centered on arthropod, and especially on insect predator-prey systems. These often make ideal laboratory subjects: the generation time is characteristically short, often enabling one to study several generations per year, and predation can be readily studied within the confines of a small cage. This allows several aspects of predator searching behavior to be investigated in isolation. Arthropod predators also present a wide range of life cycles and behavior (see Clausen, 1962, for a broad survey). Some are monophagous, others polyphagous; some attack sedentary prey, others mobile prey; some hunt by sight, others by smell; some are voracious, others require only a single prey to complete development. In addition to this diversity, there is among the arthropod predators one very large category of insects whose life cycles are in some ways rather simpler than those of other predators and hence more appropriate to the population models that are a recurrent theme of this book. These are the insect *parasitoids* (often loosely called "insect parasites") that make up about 14% of the one million or so known insect species and thus about one in ten of all species of Metazoa (Rothschild, 1965; Askew, 1971). They largely belong to two orders, the Diptera and Hymenoptera, and differ from true parasites, in the strict zoological sense, in that they almost invariably kill their hosts. Insects from almost all the orders (as well as spiders and woodlice) are subject to attack, and egg, larval, and pupal stages are more frequently parasitized than adults.

There are several important differences between the typical parasitoid and arthropod predator life cycles. In contrast to predators where usually males, females, and immature stages must locate and consume prey, it is only the adult female parasitoid that searches for hosts, and then primarily to oviposit on, in, or near to the host rather than to consume it. The behavior involved can be most sophisticated: some species use olfactory cues to locate their hosts, either responding to volatile substances from the host food-plant or from the host itself, others use vision or sound; some species paralyze their hosts before oviposition; some feed on host fluids to obtain sufficient protein for egg maturation, others feed on honeydew, nectar, or not at all; and some are efficient at detecting whether or not a host has already been parasitized, responding either to "markers" left by the previous female or to changes in the host haemolymph. Considerable variation also occurs in the number of eggs laid by a female for each host found. A single egg is laid where each host supports only one larva to complete development. Should such a host be parasitized

6

on separate occasions, it is usual for the first larva present to be the sole survivor. On the other hand, many species lay several eggs per host, or in a few cases a single egg which divides by polyembryony to give rise to numerous larvae, all of which can complete their development within the one host. When a parasitoid egg hatches, the larva either feeds from the outside of the host (as an ectoparasitoid) or from within (as an endoparasitoid). Initially, it causes little serious damage, feeding as a true parasite. However, as it approaches pupation, it begins to feed grossly on vital host organs and the host is usually killed by the time the parasitoid pupates, close to or within the host remains.

Two features of this life cycle permit important simplifications in developing population models.

(1) Since only the adult female parasitoid searches, we need seek only one set of parameters to describe the outcome of search. Parasitoid-host systems may thus be collapsed more appropriately into single age-class models. In contrast, age structure is likely to be an essential ingredient of other predator-prey models since most predators will search with different abilities during their development and probably for prey of different sizes.

(2) The number of hosts parasitized by a parasitoid population must perforce closely define the number of subsequent parasitoid progeny. This follows simply from each host parasitized by a given species tending to yield a constant number of filial parasitoids for the next generation. There will thus be a one-to-one relationship in the simplest cases where each host can support only a single parasitoid larva to maturity. Reproduction in true predators is more difficult to define, depending in part on the feeding success of the predator during development and on the several factors affecting adult female fecundity.

7

It is to insect parasitoids, therefore, that we should look for the most detailed correspondence between observed relationships and the predictions of simple population models such as those of Lotka and Volterra, Nicholson and Bailey, and their descendents. Experiments with arthropod predators remain invaluable in studying the components of searching behavior and predator reproduction and in pointing the way to the development of models more appropriate to true predator-prey interactions. Despite the differences between parasitoids and "true" predators, it is convenient in this book to write generally of arthropod "predators" and only to make the distinction between parasitoids and predators where some difference is to be emphasized.

Taken as a whole, the models in this study are presented with the object of showing how many of the components of predation may be simply modeled in order to reveal their effects on the overall dynamics of the interacting populations. In this way, we hope to develop a picture of the really important components of predation which influence both natural interactions and the outcome of biological control programs. Wherever possible, especial emphasis is placed on displaying the data that has led to the evolution of these models. This accumulation of results, largely from laboratory experiments, provides an empirical base that is probably more substantial than in any other area of theoretical ecology.

Predator-prey models have traditionally been couched in one of two mathematical formats—as differential or difference equations. Differential equations are appropriate to life cycles where generations overlap completely and birth and death processes are continuous, an ideal most frequently met in such equable conditions as in a tropical rain forest, or a heated greenhouse. Throughout this book the emphasis will be at the other extreme, with difference equations. These deal with population changes over discrete time units (often a generation interval) and therefore have the merit of including some of the time delays which figure so prominently in the real world, whether due to discrete breeding or to some delay in resource recovery time (May, Conway, Hassell, and Southwood, 1974). Time delays can, of course, be included explicitly in differential models, but this is rarely done due to the intractable mathematics that normally results (but see McMurtrie, 1975). Difference equations are therefore most appropriate to populations with quite distinct generations as frequently occur among insects in the temperate regions where diapause during winter months is common. Recently, Auslander, Oster, and Huffaker (1974) have shown that difference models may also be appropriate to systems with more complex age-class interactions, but which by virtue of their internal dynamics approach a state where the overlap in generations is reduced.

We commence with a generalized predator-prey model of the form

$$N_{t+1} = \lambda N_t f(N_t, P_t) \qquad (1.1a)$$
$$P_{t+1} = cN_t[1 - f(N_t, P_t)], \qquad (1.1b)$$

where N_t, N_{t+1} and P_t, P_{t+1} are respectively the prey and predator populations in successive generations, λ is the net rate of increase of the prey per generation—either assumed to be constant or itself some function of prey density ($\lambda = F(N_t)$)—and c is the average number of predator progeny produced per prey attacked (which we will assume to be one). The function f defines the survival of prey from predation and therefore contains all assumptions made about predator searching behavior.

The structure of this model is particularly appropriate to insect parasitoids on several counts. In the first place, it conforms well to the simplifications in the parasitoid life cycle already outlined. Thus the lack of age structure in the model applies well to parasitoids where only the adult females search for a particular host stage. In addition, the way in which the number of prey attacked defines reproduction in equation (1.1b), is appropriate to parasitoids, but less so to true predators, whose reproduction is better represented by the general equation of Beddington, Free, and Lawton (1976); namely,

$$P_{t+1} = P_t Q(N_t, P_t), \qquad (1.2)$$

where the function Q defines the per capita rate of increase of the predator as a function of prey and predator densities.

The ordering of the chapters in this book follows the logical framework of moving from the simplest to the more detailed predator-prey systems, starting first with a simple specific form for equation (1.1). Chapter 2 thus commences with the model of Nicholson and Bailey (1935)—its assumptions, properties, and relationship to the Lotka-Volterra model. This detailed explanation is necessary because it provides the foundation upon which many later refinements will be built. Such refinements fall within four broad categories. There are those factors affecting (1) the prey's rate of increase, (2) the death rate of the prey due to predation, and (3) the rate of increase of the predator population. Finally, (4) we can abandon the two-species system in favor of more complex communities with more than one prey or predator species. All of these elaborations will be considered, but some in more detail than others.

A density dependent rate of increase on the part of the prey ($\lambda = F(N_t)$) is included in Chapter 2 in relation to the Nicholson-Bailey model and later in other models in Chapters 4, 6, and 7. The prey's death rate due to predation is examined in Chapters 3, 4, and 5; in particular, the influence of prey density is considered in Chapter 3, of non-random search for a spatially heterogeneous prey in Chapter 4, and of predator interference and non-random search in Chapter 5. These are all important in shaping the function $f(N_t, P_t)$ from equation (1.1) and hence to the properties of the interaction as a whole. In Chapter 6, some factors affecting the predator's rate of increase are presented; in particular, we consider the dependence of survival, developmental rate, and adult fecundity on prey density. These determine the form of $Q(N_t, P_t)$ in equation (1.2) and thus draw us away from parasitoids and a little closer to true predator-prey interactions.

A single predator-single prey interaction in a closed, coupled system is a convenient starting point and appropriate to several interactions where the predator is specific to and synchronized with its prey. Most systems, however, are considerably more complex: either the predators attack more than one prey species or several predator species attack a single prey. Some first steps in exploring such interactions are outlined in Chapters 7 and 8. Chapter 7 considers preference and switching by polyphagous predators and how these affect the competitive outcome between prey species. Chapter 8 looks at some conditions for more than one predator species to coexist on a single prey and ends with a brief glimpse of another trophic level—the hyperparasitoid that attacks the parasitoid that attacks the host.

Finally, in Chapter 9 we turn to biological pest control and how such models can contribute to a theoretical basis for biological control practices.

In all these chapters, the mathematical details of the models have been largely omitted. They would clutter the text and obscure the insights that emerge from the models. The reader interested in the finer details is referred to the original papers or, occasionally, to one of the three mathematical appendices toward the end of the book.

CHAPTER TWO

A Basic Model

In developing predator-prey models, it is useful to commence with one that is very simple; this will then serve as a "control" situation against which the effects of additional refinements can be judged. The Nicholson-Bailey model (Nicholson, 1933; Nicholson and Bailey, 1935) provides such a "control". It is a difference model of the form of equation (1.1), and contains just a single parameter to describe the outcome of predator search. Nicholson formulated his basic model with insect parasitoids in mind, being fully aware of the greater complexity required of a true predator-prey model. His assumptions on searching behavior, however, apply equally well to predators; it is only in the lack of age structure and realistic predator reproduction that the model departs from a generalized predator-prey model.

Nicholson made two important assumptions on parasitoid searching behavior.

(1) *The number of encounters with hosts, N_e, by P_t parasitoids is in direct proportion to host density N_t.* Thus

$$N_e = aN_t P_t, \tag{2.1}$$

where the constant a is the probability that a given predator will encounter a given prey during its searching lifetime. Alternatively, as is clear from equation (2.1), it may be viewed as the proportion of the total hosts encountered by a parasitoid during that period. Nicholson assumed a to be a species specific characteristic which he called the "area of discovery". Its dimensions depend upon those of N_t and P_t: if they are expressed as total populations, a is dimensionless; but if expressed as numbers per unit area, then a will be in the same units of area. The precise meaning of N_e differs between parasitoids and predators. For parasitoids, a single host can be encountered several times and N_e is the total number of encounters with hosts, whether or not an egg is laid at each encounter. Predators, on the other hand, usually remove their prey as they are eaten, so preventing re-encounters; N_e now represents the number of encounters with "prey positions." Alternatively, were the consumed prey to be immediately replenished, then N_e becomes again the total encounters with prey.

Equation (2.1) defines the functional response (see Chapter 3) that is implicit in the Nicholson-Bailey model. It is a linear response where the slope of the relationship is defined by the searching efficiency constant a. Clearly, Nicholson's parasitoids are never limited by their egg supply (or appetite, if predators), nor is searching time a limiting factor in the number of hosts they can encounter.

(2) *These N_e encounters are distributed randomly among the available hosts.* Nicholson, following Thompson (1924), assumed that the probability of a particular host not being attacked is given by the zero term of the Poisson distribution, namely,

$$p_0 = \exp(-N_e/N_t). \tag{2.2}$$

From this, the number of hosts actually parasitized N_a (whether it be one or more times) is given by

$$N_a = N_t[1 - \exp(-N_e/N_t)], \tag{2.3}$$

an expression that merely serves to distribute N_e encounters randomly among the N_t prey. (A full treatment of the derivation of this equation is given in Appendix I).

Equation (2.3) provides a general model for predation or parasitism where search is random. We now need only an expression for N_e/N_t in order to obtain a specific population

13

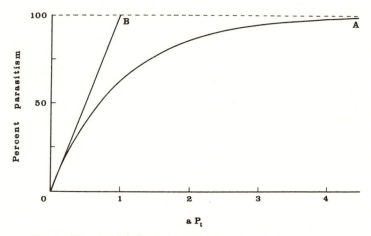

FIGURE 2.1. *(A)* Nicholson's "competition curve," obtained from equation (2.4). *(B)* The linear relationship arising if the parasitoids, rather than searching randomly, cooperate to avoid revisiting any previously searched areas (see Appendix II).

model of the form of equation (1.1). Nicholson assumed that $N_e/N_t = aP_t$, in which case we have

$$N_a = N_t[1 - \exp(-aP_t)], \tag{2.4}$$

which is the mathematical basis for Nicholson's so-called "competition curve" shown in Figure 2.1*A*. Since each host parasitized was assumed to lead to one adult parasitoid in the next generation (i.e. $P_{t+1} = N_a$), the Nicholson-Bailey population model is now obtained by substituting $\exp(-aP_t)$, the fraction of hosts surviving parasitism, for $f(N_t, P_t)$ in equation (1.1):

$$\begin{aligned} N_{t+1} &= \lambda N_t \exp(-aP_t) \\ P_{t+1} &= N_t[1 - \exp(-aP_t)], \end{aligned} \tag{2.5}$$

where λ is the net rate of increase of the hosts. The equilibrium populations, N^* and P^*, of such models are simply found by setting $N_{t+1} = N_t = N^*$ and $P_{t+1} = P_t = P^*$ which,

14

for (2.5) above, gives

$$N* = \frac{\lambda \log_e \lambda}{(\lambda - 1)a}$$

$$P* = \frac{\log_e \lambda}{a}$$

(2.6)

and from which the equilibrium surface in Figure 2.2 has been obtained.

In the real world, with its environmental vagaries, such an equilibrium solution is meaningful only if the system tends to return to these levels following a disturbance. It is widely known that the Nicholson-Bailey model has an unstable equilibrium and yields diverging oscillations when

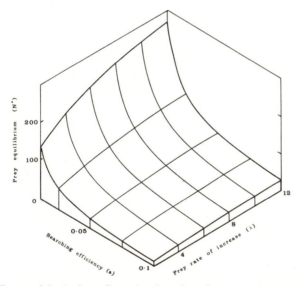

FIGURE 2.2. A three-dimensional surface from equation (2.6), showing the dependence of the equilibrium host or prey population $N*$ on both the parasitoid searching efficiency a and the host rate of increase λ. High values of a and low values of λ both promote low equilibrium populations (from Hassell, 1977).

either population is perturbed (Figure 2.3), a feature which is readily confirmed by a linearized stability analysis as outlined by Hassell and May (1973), and in Appendix II. Although this analysis is appropriate only in the neighborhood of the equilibrium, extensive numerical simulations strongly suggest that in this case the neighborhood properties do indeed also characterize the global properties following very large perturbations.

The outcome in Figure 2.3 is at marked variance with that of the Lotka-Volterra model, which is characterized by neutrally stable cycles whose amplitude depends on the initial population densities. But, as May (1973, 1975a) has pointed out, this seemingly fundamental difference between the two models has little to do with assumptions

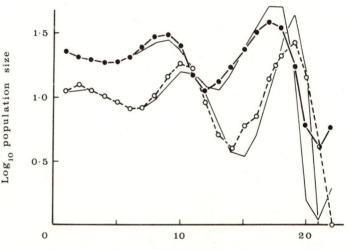

FIGURE 2.3. Population fluctuations from an interaction between the greenhouse whitefly, *Trialeurodes vaporariorum* (●), and its chalcid parasitoid, *Encarsia formosa* (○). The thin lines show the outcome from a Nicholson-Bailey model where $a = 0.068$ (the mean value over the 22 generations) and $\lambda = 2$ (the value imposed by the experimental design) (after Burnett, 1958).

16

about predator behavior; it stems from one model being framed in differential equations with no time lags, and the other in difference equations with a one-generation time lag. Redefining the Lotka-Volterra model in difference equations yields a model that now also exhibits diverging oscillations and is qualitatively very similar to that of Nicholson and Bailey (see Appendix III).

It is ironic that Nicholson, whose central interest was in the factors that might stabilize populations, should develop a general predator-prey model that is unstable. Only under certain laboratory conditions have such unstable interactions been observed (e.g. Gause, 1934; Huffaker, 1958; Burnett, 1958b). In particular, Burnett's experiment with the greenhouse whitefly, *Trialeurodes vaporariorum*, and its chalcid parasitoid, *Encarsia formosa*, provides the closest correspondence with Nicholson's model (Figure 2.3). Host and parasitoid were allowed to interact over many generations, but each "generation" was limited to 48 hours, after which the surviving hosts were doubled ($\lambda = 2$) to provide N_{t+1}, and the parasitized hosts were replaced by an equal number of parasitoids to give P_{t+1}. Burnett was therefore able to calculate the value of a for each generation using equation (2.4), rearranged to give

$$a = \frac{1}{P_t} \log_e \left[\frac{N_t}{N_t - N_a} \right], \tag{2.7}$$

and then to use the average value in his simulation shown in Figure 2.3. His experimental design obviously forced his system to be of the form of equation (1.1). The only point at issue was whether the use of the constant a would adequately represent parasitoid search. This it seemed to do.

Nicholson was, of course, fully aware that increasing oscillations are not a normal feature of natural interactions, and suggested that populations tend to exist in spatially separated fragments, with a tendency for instability to occur in each. Local extinction could therefore well occur, and immigration and emigration of predators and prey would become essential processes in maintaining the interaction as a whole. Such a picture may be quite appropriate to a few interactions. The supposed "hide-and-seek" of the moth, *Cactoblastis cactorum*, and its food-plant, the prickly pear, has often been quoted in this context (Nicholson, 1947; Andrewartha and Birch, 1954), but even this example is now reinterpreted, with interference between the clumped *Cactoblastis* larvae being thought to provide an essential stabilizing mechanism (Birch, 1971; Monro, 1975; Caughley, 1976).

It seems clear that Nicholson's view does not provide a convincing general explanation for the persistence of coupled predator-prey interactions in the real world. Instead, we must look to other features of the interactions that may be contributing to stability. One class of these arises from the predators' searching behavior and will be examined in detail in subsequent chapters. Thus, we shall see that the searching efficiency a, rather than being a species specific constant, is a function of prey density (Chapter 3), a function of prey and predator distributions (Chapter 4), and a function of predator densities (Chapter 5), all of which affect the stability properties of the populations. There remains, however, a further significant means by which interactions may be rendered stable: the prey may suffer some density dependent effect, due perhaps to resource limitation, so that their effective rate of increase λ also becomes a function of their density. The rest of this chapter is therefore devoted to the ways that density dependent prey growth rates can influence predator-prey dynamics. But first, we should note the properties of some single species models with density dependence.

SINGLE SPECIES MODELS

A variety of discrete models to describe density dependent population growth in the face of limited resources have been proposed, of the form

$$N_{t+1} = N_t F(N_t), \tag{2.8}$$

where $F(N_t)$ is some non-linear function of N_t. These models and their interesting dynamical properties have been reviewed in May, Conway, Hassell, and Southwood (1974), May (1974, 1975a, b, 1976a, b) and May and Oster (1976). Here we shall mention but two of these (i and ii below), both of which figure in this and subsequent chapters.

(i) $$F(N_t) = \lambda(1 + \theta N_t)^{-b}, \tag{2.9}$$

where θ and b are constants. This model is due to Hassell (1975) and was developed as a convenient means of describing situations ranging between the extremes of "scramble" ($b \rightarrow \infty$) and "contest" ($b = 1$), as shown in Figure 2.4. The constant θ is related to the inflection of the curve and hence to the population size above which the effects of competition become increasingly marked.

(ii) $$F(N_t) = \exp[r(1 - N_t/K)], \tag{2.10a}$$

where r is the potential growth rate per generation ($r = \log_e \lambda$) and K is the equilibrium or "carrying capacity" when $r > 0$. By identifying e^r with λ and r/K with g, this can be written in another familiar form, namely,

$$F(N_t) = \lambda \exp(- gN_t). \tag{2.10b}$$

This model has an extensive pedigree going back to Moran (1950) and thence to Ricker (1954), Cook (1965), and May (1974). The exponential increase in mortality rate with increases in population density makes the model particularly suitable for situations where the population is regulated at

19

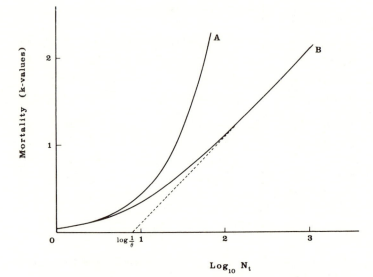

FIGURE 2.4. Examples of density dependent relationships from equation (2.9). The mortality is expressed in k-values (Haldane, 1949; Varley and Gradwell, 1960): $k = \log_{10}(N_t/N_s) = b \log_{10}(1 + \theta N_t)$, where N_s are the survivors from competition. Curve A ($b = 10$; $\theta = 0.01$): this approaches a "scramble" situation, where the resource tends to be equally divided amongst the competitors. Ideally, the mortality would rise abruptly above some threshold (i.e. $N_t = 1/\theta$) at which there is just insufficient resource to maintain any individual. Curve B ($b = 1$; $\theta = 0.12$): this represents a "contest" situation where the resource is unequally divided among the competitors; some individuals get all they require while others have insufficient for survival or reproduction (from Hassell, 1976).

high densities by epidemics (May, 1976a) or where scramble competition occurs (Figure 2.4). Indeed, the model is a special case of equation (2.9), obtained when $b \rightarrow \infty$ (with θb kept constant).

The most detailed treatment of the dynamical behavior of such time-delayed difference models is given in May and Oster (1976). For the purposes of this chapter, we need merely be aware that there can be monotonic or oscillatory approaches to an equilibrium, stable limit cycle behavior, or "chaos", as illustrated in Figure 1 of May (1975b). Being a three-parameter model, equation (2.9) is well suited to describe a range of examples where density dependence has been found from life table studies on natural populations of insects with more or less discrete generations. This is the origin of the 24 points in Figure 2.5 (solid circles); the hollow circles come from laboratory experiments. A full description of this analysis is given in Hassell, Lawton, and May (1976), who conclude, albeit with plentiful caveats,

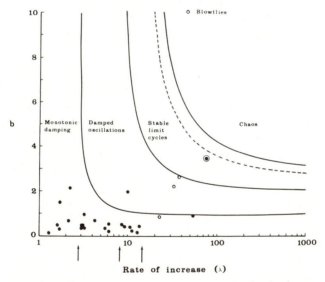

FIGURE 2.5. The stability boundaries between the density dependent parameter b and the population rate of increase λ, from equation (2.9). The parameter θ has no effect on the stability properties of the model. The solid lines separate the regions of monotonic and oscillatory damping, stable limit cycles and chaos, while the broken line indicates where two-point limit cycles give way to higher order cycles. The solid circles come from analyses of life table data and the hollow circles from laboratory experiments (the ringed point in the limit cycle region is for the Colorado potato beetle discussed in the text). The arrows mark the minimum values of λ for the different stability regions, obtained from equation (2.10) in which the stability properties hinge solely upon λ (after Hassell, Lawton, and May (1976), in which details of the individual points are given).

that high order limit cycles and chaos are unlikely to occur in populations of arthropods with fairly discrete generations. Only one point among natural populations, that of the Colorado potato beetle, *Leptinotarsa decemlineata* (ringed in Figure 2.5), appeared to be locally unstable, which is in accord with the observed population fluctuations in Ontario (Harcourt, 1971). Among laboratory experiments, only Nicholson's (1954) blowflies lie firmly in the chaotic region, as also found by Oster and Guckenheimer (1976). This is not surprising in view of the large of λ and the very heavy mortality due to scramble competition ($b \rightarrow \infty$).

The stability properties using equation (2.10) are broadly similar to those above, but now hinge solely upon the value of λ as shown by the arrows along the λ-axis of Figure 2.5. Although it is less able to describe the range of density dependent relationships found in the real world, equation (2.10) is particularly useful as a description of density dependence in predator-prey models, since there is now one less parameter with which to contend. It will therefore be used throughout this book—with the Nicholson-Bailey model in the next section and with more complex models later on.

DENSITY DEPENDENCE IN THE NICHOLSON-BAILEY MODEL

Intuitively, we would expect that the inclusion of a density dependent prey growth rate would have a marked effect on the dynamics of a predator-prey model. This was demonstrated by Varley and Gradwell (1963) when they showed by simulation that the density dependent pupal mortality of the winter moth *(Operophtera brumata)* could stabilize a Nicholson-Bailey interaction between the winter moth and one of its parasitoids (Figure 2.6). The model

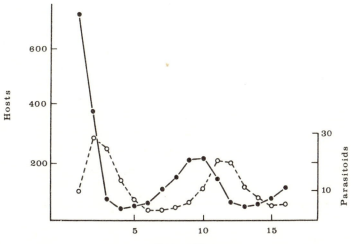

FIGURE 2.6. Host (●) and parasitoid (○) population fluctuations obtained from equation (2.11) where λ = 10, a = 0.063 m^2, and b = 0.35. These parameter values are based on life table studies on the winter moth and its parasitoids (after Varley and Gradwell, 1963).

they employed took the form:

$$N_{t+1} = \lambda N_t^{(1-b)} \exp(-aP_t)$$
$$P_{t+1} = N_t^{(1-b)}[1 - \exp(-aP_t)], \qquad (2.11)$$

where b is a constant akin to that in equation (2.9). This, however, is not well suited as a general model since it has the disconcerting feature of the prey suddenly exhibiting *positive* feedback below some critical population density (May, Conway, Hassell, and Southwood, 1974; Hassell, 1975).

More recently a much more detailed treatment of the effects of density dependence in predator-prey models has been given by Beddington, Free, and Lawton (1975). They

23

incorporate the function (2.10a) within the Nicholson-Bailey model to give

$$N_{t+1} = N_t \exp[r(1 - N_t/K) - aP_t]$$
$$P_{t+1} = N_t[1 - \exp(-aP_t)]. \tag{2.12}$$

The local stability boundaries for this model (see Figure 2.7) show that in contrast to the unstable Nicholson-Bailey model, a stable predator-prey equilibrium may now exist, with stability hinging upon two parameters. These are the prey reproductive rate ($r = \log_e \lambda$) and a term q, defined as N^*/K, which is thus a measure of the extent to which the prey equilibrium N^* is depressed by predation below its

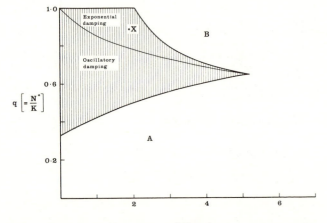

FIGURE 2.7. Stability boundaries for the predator-prey model (2.12). The parameter q is a measure of the extent to which the prey equilibrium N^* is depressed by predation below its carrying capacity K. The equilibrium point is stable within the hatched area only. Outside this domain, limit cycle and chaotic behavior occur. In region A, the fluctuations are of the order of those from natural populations, while in region B they are so great that the predators would effectively become extinct. Point X represents an arbitrary position within the stable area to illustrate the "paradox of enrichment" discussed in the text (after Beddington, Free, and Lawton, 1975).

24

carrying capacity K. Hence, for a given value of λ and K, the depression q depends solely on the predator efficiency term a; as a increases from very low values, q moves from almost unity $(N^* \simeq K)$ toward zero $(N^* \ll K)$. Qualitatively the same result has been discussed by Southwood and Comins (1976; see their Figure 5a).

Figure 2.7 shows that for realistic values of λ, there is an appreciable area of parameter space in which a stable equilibrium exists. Thus, as long as q is not less than about 0.4, the density dependence in the prey growth rate could be sufficient to counteract the instability due to the Nicholson-Bailey interaction, given appropriate values of r. Outside this stable domain, limit cycle and chaotic behavior occur, as elegantly shown by the trajectories in Beddington, Free, and Lawton (1975); but only within region A is the range of population fluctuations quite consistent with those observed in natural populations (Table 3 of Hassell, Lawton, and May, 1976). In region B, the fluctuations are so great that to all intents and purposes the predator becomes extinct.

As May (1976c) has noted, models such as (2.12) point clearly to the mechanism underlying Rosenzweig's (1971) "paradox of enrichment," whereby increasing the level of the prey's carrying capacity K leads to reduced stability, and eventually to stable limit cycle behavior. Hence if K is progressively increased but N^* kept more or less constant, the overall prey dynamics will become less and less influenced by the prey density dependence and increasingly dominated by the predator. A point such as X in Figure 2.7 will thus move downward through the region of exponential and oscillatory damping and thence to the unstable region A.

We have seen in this chapter how density dependent prey growth rates can render the basic Nicholson-Bailey model quite stable. While such density dependence must always

25

apply to populations faced with limited resources, predator-prey equilibria will often occur at levels where this effect is not at all marked (as in the successful cases of biological control, discussed in Chapter 9). We should therefore also look to the predators themselves as a means of stabilizing interactions. The next three chapters are thus devoted to more reasonable functions for predation, concentrating in particular on some crucial aspects of predator searching behavior. We shall see that with real animals the searching efficiency term a is by no means constant, nor is search random. The more realistic models that result will show that predator-prey interactions can of themselves be quite stable without any help from their prey's dynamics.

SUMMARY

The Nicholson-Bailey model provides a convenient starting point for exploring the dynamics of coupled predator-prey models framed in difference equations. It is the basis upon which many of the elaborations of later chapters will rest. The model makes two fundamental assumptions about the searching of predators: (1) that they encounter prey in direct proportion to the prey density (implying a constant searching efficiency a and an unlimited appetite, or egg supply for parasitoids), and (2) that these encounters are distributed randomly among the available prey.

The unstable properties of this model prompt the question of why real interactions generally prove to be stable, even in several simple laboratory systems. There are two likely explanations: (1) that to encapsulate predator searching behavior within a single constant is a gross simplification, and more realistic assumptions of the ways predators search would yield stable models, and (2) that a density dependent reproductive rate (due perhaps to competition for a limiting resource) would also contribute markedly to stability. A suitable expression for this density dependence, and the way that it can stabilize the Nicholson-Bailey model, conclude this chapter, leaving the elaborations of predator searching behavior to follow in succeeding ones.

CHAPTER THREE

Functional Responses

The extent to which a prey population suffers predation hinges both upon the number of predators present and on their ability to find and consume (or parasitize) their prey. It is therefore logical, in studying the dynamics of predation, to make some distinction between those factors affecting predator abundance and those affecting predator searching efficiency. The conventional way of doing this is to recognize two predator responses: the *functional* and *numerical* responses, a terminology that originated with Solomon (1949), but was exploited later by Holling (e.g. 1959a, b, 1961, 1965, 1966). A functional response was defined in terms of the relationship between the number of prey consumed per predator and prey density, and the numerical response as the relationship between the number of predators and prey density.

A broader classification, discussed by Hassell, Lawton, and Beddington (1976) and Beddington, Hassell, and Lawton (1976), also distinguishes between two aspects of predation: some factors affect the *prey death rate* and others the *predator rate of increase*. This distinction parallels to some extent that between functional and numerical responses, but differs in prey density being viewed as only one of the independent variables affecting prey consumption and predator numbers. Searching efficiency, for instance, is influenced by the vagaries of climate (see Klomp, 1959, for a good example), the density of prey (the functional response), the prey distribution, the density of predators, and any alternative prey or competing predator species. Of these, the functional response to prey density in a homogeneous environment is singled out in this chapter. Later chapters deal with the response to a spatially heterogeneous prey (Chapters 4 and 5); the response to predator density (Chapter 5); and how these are reflected in the developmental rate, survival, and fecundity of the predators, which in combination determine the predator rate of increase (Chapter 6). All this is in the context of a specific predator attacking a given prey species. This assumption is relaxed in Chapters 7 and 8; Chapter 7 deals with multi-prey systems and Chapter 8 with systems involving more than one predator species.

Intuitively, we would expect a functional response to take the form of an increasing number of prey eaten per predator as prey density increases, at least up to some limiting value representing maximum prey consumption within the prescribed time. Holling (1959a) considered three such responses, illustrated in Figure 3.1: *type I* where the response rises linearly to a plateau; *type II* where the response rises at a continually decreasing rate, and *type III* where the response is sigmoid. The aim in this chapter is to identify the functional responses that are particularly appropriate to arthropod systems, to model these in a general way, and then consider their effects on the outcome of predator-prey

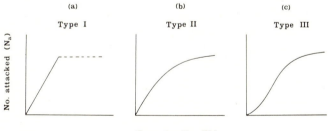

FIGURE 3.1. The three types of functional response to prey density proposed by Holling (1959a). The type I response is discussed in the text as a linear relationship and hence the upper plateau ignored.

29

models. To do this, we shall consider each of the three types of response in turn.

TYPE I RESPONSES

These responses are supposedly typical of aquatic filter-feeding invertebrates that waft in plankton in direct proportion to its surrounding abundance. Such feeding tends to cease abruptly when the animal is satiated, giving rise to the discontinuous plateau shown in Figure 3.1a and hence to what is really just a special case of the type II response to be considered in the next section. In this section, we shall examine briefly a linear functional response of the kind implicit in both the Nicholson-Bailey and Lotka-Volterra models. We are assuming, therefore, that our predators do not become satiated, nor our parasitoids run out of eggs.

Assuming a constant time T available for search, such linear responses may be described by

$$N_e/P_t = a'TN_t, \qquad (3.1)$$

where N_e is the number of encounters with prey and a' is the instantaneous search rate as evident from rearranging equation (3.1). The combination $a'T$ is thus identical to Nicholson's area of discovery a in equation (2.1), which in his models merely represents the search rate throughout the parasitoid or predator's searching lifetime (i.e. $T = 1$ generation).

In the case of parasitoids, equation (3.1) describes the number of encounters with hosts throughout the course of the interaction T. Alternatively, it can be viewed as the number of encounters with prey if the prey density N_t remains effectively constant during T, either by each prey being replaced as soon as eaten, or by the predators searching systematically and so avoiding covering any area more than once (Rogers, 1972). The subtle distinction between

N_e, the number of encounters with hosts or prey, and N_a, the number of hosts parasitized or prey eaten, is an important one; N_e can exceed N_t, while N_a can never do so.

To express equation (3.1) in terms of N_a, we simply substitute it in equation (2.3) to give

$$N_a = N_t[1 - \exp(- a'TP_t)], \qquad (3.2)$$

the formal derivation of which is described in some detail in Appendix I. In this way, the basic Nicholson-Bailey equations (2.4) and (2.5) are reobtained, whose unstable properties have already been considered in Chapter 2.

For most predators, linear functional responses are an inadequate description of their response to prey density, since the searching time cannot be a constant. Some time will inevitably be expended over each prey item eaten, and hence searching time progressively reduced as more prey are consumed. This is central to the type II responses discussed in the next section.

TYPE II RESPONSES

A variety of type II functional responses for arthropod predators and parasitoids is shown in Figure 3.2, all showing a decelerating rise to an upper asymptote. It was Holling (1959b) who first pinpointed the essential biological ingredient that distinguishes these type II from type I responses: that the act of quelling, killing, and eating a prey, and then perhaps cleaning and resting, are all time-consuming activities (collectively called the "handling time") which reduce the time available for search. We must distinguish, therefore, between the total time initially available for search T, and the actual searching time T_s, which depends on the number of prey encountered and is given by

$$T_s = T - T_h(N_e/P_t), \qquad (3.3)$$

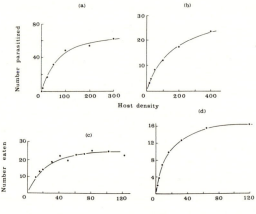

FIGURE 3.2. Some typical type II functional responses for insect parasitoids (a, b) and insect predators (c, d). The parameters a' and T_h were estimated from equation (3.5) or (3.6) using a non-linear least squares technique, and are expressed in units of "days^{-1}" and days respectively. (a) *Nasonia vitripennis* (40 females) parasitizing *Musca domestica* pupae (DeBach and Smith, 1941b), $a' = 0.027$, $T_h = 0.52$. (b) *Dahlbominus fuscipennis* parasitizing *Neodiprion sertifer* cocoons (Burnett, 1956), $a' = 0.252$, $T_h = 0.037$. (c) Final instar damselfly larvae *(Ischnura elegans)* feeding on *Daphnia magna* (Thompson, 1975), $a' = 1.38$, $T_h = 0.032$. (d) Second instar coccinellids *(Harmonia axyridis)* feeding on *Aphis craccivora* (Mogi, 1969), $a' = 1.87$, $T_h = 0.056$.

where T_h is the handling time. Substituting this in equation (3.1) now gives us the familiar "disc equation" of Holling (1959b); namely,

$$\frac{N_e}{P_t} = a' \left[T - T_h \frac{N_e}{P_t} \right] N_t \qquad (3.4a)$$

or

$$\frac{N_e}{P_t} = \frac{a'TN_t}{1 + a'T_hN_t}, \qquad (3.4b)$$

with T/T_h defining the maximum number of prey that can be eaten and a' determining how rapidly the curve approaches this upper asymptote.

This equation is often used to describe data such as in Figure 3.2, with a' and T_h being obtained from a linear regression of $N_e/(N_t P_t)$ against N_e/P_t as explained in Holling (1959b). This technique, however, is inappropriate and will yield misleading estimates of a' and T_h, unless it is really the encounters with a constant prey density that are being scored rather than the numbers of prey eaten or hosts parasitized. To overcome this problem, we once more require the functional response to be expressed in terms of N_a rather than N_e, but now there is the additional complexity of having to distinguish between parasitoids and predators. The important difference here is that a parasitized host, unlike a completely consumed prey, remains exposed to further encounters by searching parasitoids. Most parasitoid-host models make the assumption that such re-encounters will lead to exactly the same period of handling time as the first encounter, whether or not a further egg is laid. Consequently, altogether more time will be spent in handling by a parasitoid than by an equivalent predator. To a first approximation, this assumption will be adequate for many parasitoids, although the length of the handling time is likely to be reduced where superparasitism is avoided, as found by Cook (1977) for the ichneumonid *Nemeritis canescens*. The behavioral refinement, that a parasitized host is only cursorily examined at a subsequent time due to some distinguishing mark or odor left by the previous female (e.g. *Trichogramma evanescens*, Salt, 1937), is probably less common and approaches the predator situation where, of course, re-encounters do not figure at all.

The appropriate "attack equations" that arise from the disc equation, for both parasitoids and predators, were first given by Royama (1971a) and Rogers (1972):

$$N_a = N_t \left[1 - \exp \left\{ - \frac{a'TP_t}{1 + a'T_h N_t} \right\} \right]$$

(for parasitoids) (3.5)

$$N_a = N_t \left[1 - \exp \left\{ -a'P_t \left(T - T_h \frac{N_a}{P_t} \right) \right\} \right]$$

(for predators)   (3.6)

the derivations of which are laid out in Appendix I. Notice that the Nicholson-Bailey equation (3.2) is now just a special case of either (3.5) or (3.6) when $T_h = 0$. As T_h increases relative to T, the difference between parasitoids and predators becomes more marked, with the parasitoid having less time than the predator available for search.

Each set of data in Figure 3.2 has been described by one of these models. The estimates of a' and T_h have come from a standard non-linear least squares technique applied directly to the untransformed data. The much simpler alternative proposed by Rogers (1972), of a linear regression applied to the transformed data, is unfortunately fraught with statistical problems and hence prone to yield biased estimates of the parameters. This has been fully discussed by Cock (1977). We should note that estimates of T_h obtained in this way from experimental data are likely to yield values that do not conform at all well to direct observations of the time that the predator or parasitoid spends visibly dealing with each prey. This arises because the estimated T_h also includes periods of non-searching activity induced, for example, by satiation in the case of predators or egg depletion in the case of parasitoids. The difference between the two estimates will be most marked where the predators or parasitoids have a maximum attack rate, or upper asymptote of the response, that is primarily determined by satiation, or the rate of egg maturation, rather than by the actual handling time.

Despite the good performance of these models in describing data, it is unlikely that both a' and T_h are indeed constant and independent of prey density and feeding rate. Each is a function of several components which have been analyzed in detail by Holling (1965, 1966). Thus the attack rate a' is likely to be a function of

(1) the maximum distance at which the predator can initiate an attack on a prey (the predator's reactive distance);
(2) the speed of movement of predator and prey; and
(3) the proportion of attacks that are successful.

The handling time T_h will depend upon

(1) the time spent pursuing and subduing an individual prey;
(2) the time spent eating each prey, or parasitizing each host; and
(3) any time spent resting or cleaning as a result of feeding by predators, or ovipositing by parasitoids.

In many predators, one or more of these components of a' and T_h are known to vary with prey density, predator feeding rate, or the time elapsed since the last meal (all of which are likely to be interrelated). Information on this is plentiful in the literature (see Hassell, Lawton, and Beddington, 1976, for a review), and includes the effects of a predator's speed of movement (Dixon, 1959; Glen, 1975), its capture success (Fransz, 1974; Nelmes, 1974), the proportion of each prey eaten (Turnbull, 1962; Haynes and Sisojevic, 1966; Glen, 1973; Cockrell, 1974; Johnson, Akre, and Crowley, 1975), and the length of time since the last meal (Nakamura, 1972; Sandness and McMurtry, 1972). It is thus very unlikely that the attack rates and handling times at the end of a functional response experiment run over any appreciable period of time will be the same as those at the beginning. An example of this is given in Figure 3.3 where the handling time of a coccinellid is seen to increase with successive feeds.

Although this point has been stressed, it remains true that the assumption of constant a' and T_h is adequate for

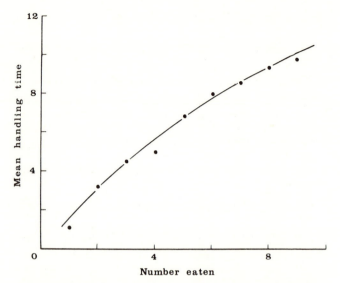

FIGURE 3.3. The results of continuous observations on the feeding of *Coccinella septempunctata* on cabbage aphids, *Brevicoryne brassicae*, under laboratory conditions. The running mean handling time in minutes is seen to increase as a function of the number of prey eaten (from Cock, 1977).

many species of predators and parasitoids, despite variations in their sub-components, and this allows much functional response data to be simply described (see Figure 3.2). Presumably, some of the changes in the sub-components tend to cancel each other out, while others have relatively little effect. Only where more complex responses occur (e.g. dome-shaped, or the sigmoid responses of the next section) is it clear that the assumption of constant a' and T_h becomes quite inadequate.

The effects of type II functional responses on stability are straightforward. Allowing the function in equation (1.1) to become

$$f(N_t, P_t) = \exp\left[-\frac{a'TP_t}{1 + a'T_h N_t}\right] \qquad (3.7)$$

36

for parasitoids, or

$$f(N_t, P_t) = \exp\left[-a'P_t \left(T - T_h \frac{N_a}{P_t} \right) \right] \qquad (3.8)$$

for predators, we now have a model that is always less stable than the corresponding Nicholson-Bailey model, which is

TABLE 3.1. Estimated values of the handling time T_h from equation (3.5) or (3.6) for a selection of parasitoids and predators. The values of T_h/T are based on conservative estimates of longevity.

Parasitoid or predator species	Host or Prey	Handling time T_h (hrs)	T_h/T	Author(s)
PARASITOIDS				
Nemeritis canescens	Ephestia cautella	0.007	<0.0001	Hassell & Rogers (1972)
Chelonus texanus	Ephestia kühniella	0.12	<0.001	Ullyett (1949a)
Dahlbominus fuscipennis	Neodiprion lecontei	0.24	<0.003	Burnett (1958)
Pleolophus basizonus	Neodiprion sertifer	0.72	<0.02	Griffiths (1969)
Dahlbominus fuscipennis	Neodiprion sertifer	0.96	<0.01	Burnett (1954)
Cryptus inornatus	Loxostege sticticalis	1.44	<0.02	Ullyett (1949b)
Nasonia vitripennis	Musca domestica	12.00*	<0.1	DeBach & Smith (1941)
PREDATORS				
Anthocoris confusus (5th instar)	Aulacorthum circumflexus	0.38	<0.001	Evans (1973)
Notonecta glauca (1st instar)	Daphnia magna	0.76	<0.005	B. H. McArdle (unpublished)
Ischnura elegans (12th instar)	Daphnia magna	0.82	<0.002	Thompson (1975)
Harmonia axyridis (2nd instar)	Aphis craccivora	1.61	<0.002	Mogi (1969)
Phytoseiulus persimilis (Adult ♀)	Tetranychus urticae	1.87	<0.005	Pruszyński (1973)

* This figure is the handling time for each host parasitized, in which many eggs are laid. The handling time per egg laid would be very roughly 0.4 hrs.

37

recovered when $T_h = 0$. This is shown formally by Hassell and May (1973) and is, of course, the direct consequence of predation being inversely density dependent. The extent of this additional instability, however, depends not on the absolute value of T_h but on its value relative to the total time available (i.e. T_h/T). Hassell and May conclude that handling time will be of only minor importance in affecting stability, as long as $T_h/T \ll 1$. The examples in Table 3.1 support such small ratios, at least for parasitoids, where species with very long handling times appear to compensate by being especially long-lived.

TYPE III RESPONSES

There are numerous statements in the literature to the effect that sigmoid (type III) functional responses are typical of vertebrate predators, while type II responses are characteristic of invertebrates. Evidence is accruing, however, that belies this notion (Murdoch and Oaten, 1975; van Lenteren and Bakker, 1976; Hassell, Lawton, and Beddington, 1977), and suggests that sigmoid responses are also widespread among arthropod predators and parasitoids. Some examples are shown in Figure 3.4. The most likely explanation for these results is that the predators are tending to search more actively as prey density rises, making one or more of the components of predator searching activity dependent on prey density. Support for this comes from Figures 3.5 and 3.6. Figure 3.5 shows the results of an experiment in which an individual of the ichneumonid parasitoid *Nemeritis canescens* was continuously observed for 30 minutes while exposed to one of a range of host densities. The time spent actively probing the medium appears as a simple increasing function of host density: as host density is reduced, an increasing proportion of the time available is spent in non-hunting activities,

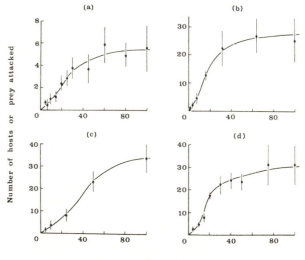

Prey density

FIGURE 3.4. Sigmoid, type III functional responses for insect predators (a, b, d) and parasitoids (c). The mean and 95% confidence
intervals are shown and all curves are drawn by eye. (a) The waterboatman, *Notonecta glauca,* feeding on *Asellus aquaticus.* (b) *Coccinella septempunctata* feeding on the aphid, *Brevicoryne brassicae*
(Cock, 1977). (c) The braconid wasp, *Aphidius uzbeckistanicus,*
parasitizing the aphid, *Hylopteroides humilis* (Dransfield, 1975). (d)
The waterboatman, *Plea atomaria,* feeding on mosquito larvae,
Aedes aegypti (A. Reeve, unpublished data) (from Hassell, Lawton,
and Beddington, 1977).

such as walking and resting on the sides of the cage. This
behavior, therefore, provides an explanation for the sigmoid response observed for *Nemeritis* by Takahashi (1969).
A similar result was obtained from experiments using adult
blowflies, *Calliphora vomitoria,* as "predators" of sugar droplets as "prey" in a simple Perspex arena (see Murdie and
Hassell, 1973, for details). The sigmoid functional response
obtained from these experiments is shown in Figure 3.6a,
and from Figure 3.6b it is clear that this again is the result
of searching activity (in this case, the time spent walking on
the floor of the arena) increasing with prey density.

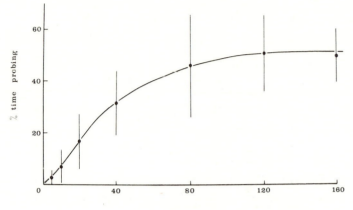

FIGURE 3.5. The relationship between the time spent probing by *Nemeritis canescens* (as a percentage of total observation time) and the density of its host, the larvae of *Plodia interpunctella* (from Hassell, Lawton, and Beddington, 1977).

In both these examples the predators hunt more actively as prey density rises, especially by spending proportionally more time searching for prey. Other, related, behavior is also possible. In the *Calliphora* experiment, for example, the speed of movement of the fly walking in the arena also appears to depend upon the frequency of encountering droplets, while for *Nemeritis* there may be a more rapid rate of probing of the host's medium at higher host densities. Neither of these particular behavioral details have been quantified.

In general, therefore, we would expect a sigmoid functional response to be explicable on the basis of a', T, or both being increasing functions of prey density. Unfortunately, most functional response experiments have been conducted without the attendant behavioral observations needed to identify these relationships. But it is still possible to show how a' or T alone would have to vary in order to give the observed response. Figure 3.7 for example, uses

40

the data from the sigmoid responses in Figures 3.4a and c and shows the relationship between the supposed value of a' and prey density. The general similarity between the way a' varies here and the observed relationships in Figure 3.5 and 3.6 is obvious.

Interestingly, the sigmoid functional responses of *Notonecta, Coccinella,* and *Aphidius* in Figure 3.4 have been converted to the typical type II responses shown in Figure 3.8, simply by making the conditions somewhat more favorable: *Notonecta* and *Coccinella* by providing larger prey

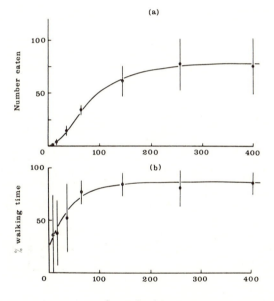

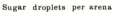

FIGURE 3.6. (a) A sigmoid functional response for adult blowflies, *Calliphora vomitoria,* feeding on evenly distributed sugar droplets in a Perspex arena (T = 30 min). (b) The relationship between searching activity of *Calliphora* and droplet density per arena. Activity is expressed as the percentage of the total observation time spent walking on the floor of the arena (corrected for periods spent feeding). Means and 95% confidence limits are shown and both curves are fitted by eye (from Hassell, Lawton, and Beddington, 1977).

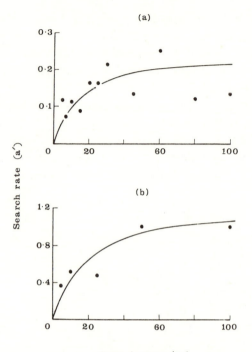

Prey density (N_t)

FIGURE 3.7. The relationship between the search rate a' and prey density N_t obtained from two of the sigmoid functional responses in Figure 3.4. In each case a constant handling time has been assumed and equation (3.5) or (3.6) then solved for a' at different N_t, knowing N_a (as observed), T_h, and T. (The search rates are defined per cage unit with $T = 1$.) (a) *Notomecta glauca* feeding on *Asellus aquaticus;* $T_h = 0.091$. (b) *Aphidius uzbeckistanicus* parasitizing *Brevicoryne brassicae;* $T_h = 0.018$ (from Hassell, Lawton, and Beddington (1977), in which the method of estimating T_h is explained).

and *Aphidius* by providing a preferred host species. Under these favorable conditions, the predators seem to be maintaining a constant search rate at even the lowest prey densities. We may conclude that sigmoid responses are to be expected whenever the "reward rate" at the lowest prey densities is insufficient to maintain this constant searching activity.

42

Results such as these suggest a direct means of obtaining models for sigmoid responses. Let us suppose that it is only a' that varies with prey density and does so in the following way, akin to the relationships in Figure 3.7:

$$a' = \frac{bN_t}{1 + cN_t},$$ (3.9)

where b and c are constants. Substituting this into equation (3.4) now gives us a sigmoid analogue of the disc equation, namely,

$$\frac{N_e}{P_t} = \frac{bN_t^2 T}{1 + cN_t + bT_h N_t^2}.$$ (3.10)

As with the type I and type II responses, it is desirable to have an expression in terms of the number of prey at-

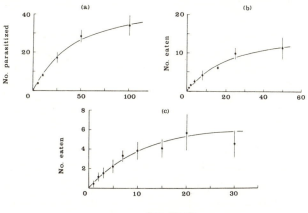

FIGURE 3.8. Type II functional responses obtained with "preferred" or large prey, for one parasitoid (a) and two predators (b, c) which were shown in Figure 3.4 to yield sigmoid responses with other prey types. a' and T_h estimated as in Figure 3.2 and expressed in units of "hours^{-1}" and "hours" respectively. (a) *Aphidius uzbeckistanicus* parasitizing *Metapolophium dirhodum* (Dransfield, 1975), $a' = 0.137$, $T_h = 0.49$. (b) *Coccinella septempunctata* feeding on large *Brevicoryne brassicae* (Cock, 1977), $a' = 0.501$, $T_h = 0.13$. (c) *Notonecta glauca* feeding on large *Asellus aquaticus*, $a' = 0.74$, $T_h = 2.64$ (after Hassell, Lawton, and Beddington, 1977).

tacked N_a, rather than in terms of N_e. The form that this takes depends upon specific assumptions: whether predators or parasitoids are being considered and, if predators, whether the instantaneous search rate a' in equation (3.9) depends upon the initial prey density or the number remaining as the interaction proceeds. For parasitoids, where the hosts remain to be subsequently re-encountered, we have

$$N_a = N_t \left[1 - \exp\left(-\frac{bTN_tP_t}{1 + cN_t + bT_hN_t^2}\right)\right]. \quad (3.11)$$

For predators, there are two alternatives. First, a' in equation (3.9) may depend on the initial prey density, in which case

$$N_a = N_t \left[1 - \exp\left\{-\frac{bN_tP_t}{1 + cN_t}\left(T - \frac{T_hN_a}{P_t}\right)\right\}\right]. \quad (3.12)$$

Alternatively, as assumed by Hassell, Lawton, and Beddington (1977), the value of a' may hinge upon the number of prey available at any moment, giving

$$N_a = N_t \left[1 - \exp\left\{-\frac{bP_t}{c}\left(T - \frac{T_hN_a}{P_t} - \frac{N_a}{bN_tP_t(N_t - N_a)}\right)\right\}\right]. \quad (3.13)$$

The detailed derivation of these models is given in Appendix I. All three yield very similar sigmoid relationships for given parameter values, especially if $N_a \ll N_t$, but that for parasitoids (3.11) is the most convenient to use since N_a may be found without iteration. For that reason equation (3.11) is adopted in the following section and in Chapter 7, where the effects of sigmoid responses on stability are discussed.

Sigmoid Responses and Population Stability

The importance of distinguishing between type II and type III responses rests on their supposedly very different contributions to stability. Only sigmoid functional responses are density dependent up to some threshold prey density, and in consequence are widely assumed to contribute to the stability of a predator-prey interaction, perhaps even providing a stable equilibrium if the prey equilibrium population falls within the density dependent, accelerating part of the response. This has been formally stated by Murdoch and Oaten (1975), Oaten and Murdoch (1975a), and Murdoch (1977) with reference to the differential Lotka-Volterra model. They showed that the system is stable provided that

$$g'(N^*) > g(N^*)/N^*, \tag{3.14}$$

where $g(N)$ is the total number of prey eaten at prey density N, and $g'(N^*)$ is its derivative evaluated at the equilibrium density N^*. While this is certainly true for a model with no time delays, there is another whole class of models where this criterion of N^* falling within the density dependent part of the response does not apply. In particular, Hassell and Comins (1978) have shown that a sigmoid functional response *alone* cannot stabilize any difference model of the form of equation (1.1).

To illustrate this, we begin with the sigmoid response equation (3.11) which, within our basic model (1.1), gives the following function for predation (or, more strictly, parasitism):

$$f(N_t, P_t) = \exp\left[-\frac{bTN_tP_t}{1 + cN_t + bT_hN_t^2}\right]. \tag{3.15}$$

The stability analysis of this model is outlined by Hassell and Comins, who show stability to be impossible regardless of the form of the functional response. The reason for this instability lies in the excessive time-delayed feedback (of one generation interval) between changes in the predator density and the level of prey mortality (aP_t) in the subsequent generation.

This analysis applies to the extreme case where the predator population is specific to and synchronized with its prey, a situation best seen in many host-parasitoid interactions. At the other extreme are predators that are generalists; their population fluctuations will tend to be unrelated to the density of a particular prey type. The properties of such a system are, of course, quite different, and now sigmoid responses can indeed be a powerful stabilizing mechanism. This is further discussed in Chapter 7.

AGE DEPENDENT RESPONSES

Predators, in contrast to parasitoids, have juveniles as well as adults that must usually search for prey whose size may also vary. They present, therefore, the considerable complication of showing functional responses that will vary markedly, depending on the stage of predator development and the size of prey taken (Murdoch, 1971; Hassell, Lawton, and Beddington, 1976). The dynamics of true predator-prey models may thus be significantly affected by incorporating age structure into population models. To do this we must first know

(1) how the parameters a' and T_h (assuming type II responses) vary among successive developmental stages of the predator when they search for the same size of prey, and

(2) how they vary within the same predator stage, but feeding on different sizes of prey. (This can also apply to parasitoids where more than one host instar—of aphids, for example—are accepted as hosts).

Few studies present us with such complete information gathered for a single predator species, notable exceptions being provided by Thompson (1975) working with the damsel fly *Ischnura elegans*, and Fernando (1977) with the predatory mite *Phytoseiulus persimilis.* Thompson obtained functional responses for a range of predator instars feeding on water fleas of different sizes, and from each response abstracted values of a' and T_h which he found to depend upon predator and prey size. These results, shown in Figure 3.9, indicate a clear tendency for a' to decline and T_h to increase as prey become larger or predators smaller. These trends are also largely supported by Fernando and by several other experiments, reviewed by Hassell, Lawton, and Beddington (1976), where predator or prey size alone have been varied. This variation in a' and T_h is readily explicable in terms of their behavioral sub-components. In general, T_h should decrease as prey size is reduced, since smaller prey will usually be easier to subdue, consume, and digest than larger prey. This has been well shown by Dixon (1959) for the coccinellid *Adalia decempunctata* feeding on nettle aphids of different size. Similarly, a' will often tend to increase as the predator grows, since larger predators are likely to move more rapidly, detect prey from greater dis-

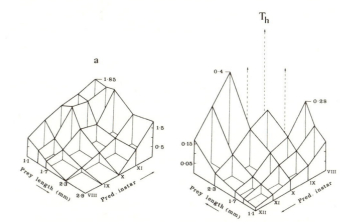

FIGURE 3.9. The effect of both predator and prey size on a' and T_h, measured under standard conditions for the damselfly, *Ischnura elegans,* feeding on *Daphnia magna* (from Thompson, 1975).

tances, and make a higher proportion of successful attacks. Examples come from the work of Dixon (1959), Wratten (1973), Brown (1974), and Glen (1975), for a variety of coccinellids and a mirid bug. In addition to these inter age-class differences, we should once more note that a' and T_h are unlikely to be invariant within a single predator age-class. At any one time, both parameters are likely to be affected by the level of hunger (in predators), or the egg complement (in parasitoids). To unravel this fully, functional response experiments are needed throughout the searching lifetime of the adult parasitoid or particular predator stage.

It is clear from these results that although a *single* functional response can provide an adequate description of the behavior of many parasitoids in relation to prey density, this is no longer true of the behavior of predators. Furthermore, since predators will often be faced at the same time with prey of different sizes, it will not be sufficient merely to predict the age structure of both populations prior to employing a series of functional responses based on such results as in Figure 3.9. We still need to know how a predator will select from a mixture of prey stages. Perhaps the problem will prove to be essentially the same as that involving predation on different prey species, as discussed in Chapter 7. There is, however, little detailed information upon which to develop this theme, and we can do no more than conjecture that such age-class effects are likely to be very important in predator-prey dynamics.

SUMMARY

The functional response to prey density is a crucial component of any predator-prey model. In this chapter, the three basic types of functional response (linear, convex, and sigmoid) are reviewed and their dynamic effects noted.

Their detailed derivation, however, is largely left to Appendix I. The linear response is implicit in both the Nicholson-Bailey and Lotka-Volterra models. The convex, type II response differs from this only in that a finite amount of time is spent handling each prey following capture. Finally, the sigmoid, type III response may be generated by assuming that either the instantaneous search rate a' or the available time T is a function of prey density. None of these responses can stabilize a coupled predator-prey system with discrete generations, although the sigmoid responses can do so if time delays are absent or the predator population is uncoupled from that of its prey.

One reason why parasitoids are such convenient experimental subjects is that there is just a single searching stage, the adult female. Predators, in contrast, present the formidable complication of also seeking prey while juveniles. Their search rates and handling times therefore vary with their stage of development and the size of prey taken. This complexity poses considerable difficulties for the modeling of predator-prey, rather than parasitoid-host, interactions.

CHAPTER FOUR

Non-Random Search

The predators in Chapters 2 and 3 searched at random for prey whose distribution was unspecified. It was as if the predators and prey dwelt in a homogeneous world. Of course, most prey populations under natural conditions exhibit a clumped distribution among units of their habitat, and within this framework, random search implies that the same number of predators spend on average the same period of time in each prey area. In this way, the searching rate remains independent of prey and predator densities. This is a convenient assumption to make mathematically, since the zero term of the Poisson distribution becomes the basis for population models, but it is a poor one biologically. Prudent predators will tend to spend more of their searching time where prey are abundant rather than scarce and hence be at a considerable selective advantage. Some examples of this are shown in Figure 4.1.

In this chapter, some means by which predators tend to aggregate in regions of high prey density are outlined; this is followed by a more detailed treatment of the effects of such non-random search on population stability. The population models are of three kinds: (1) a relatively simple model with arbitrary prey distributions and a simple predator aggregation function, (2) a more complex sequel with more realistic prey and predator distributions, and finally (3) the simplest of them all, a one-parameter characterization of the outcome of predator aggregation which captures several of the essential dynamical features of the more complex models, although containing very little biological detail.

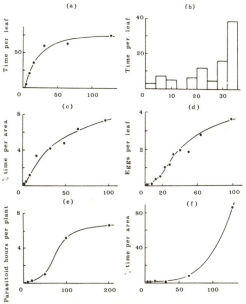

Hosts or prey per patch

FIGURE 4.1. Some examples of aggregative responses by predators and parasitoids. (All curves are fitted by eye). (a) Searching time of the braconid wasp, *Diaeretiella rapae,* at different densities of its aphid host, *Brevicoryne brassicae,* per cabbage leaf (Akinlosotu, 1973). (b) As (a), but using the larvae of the coccinellid, *Coccinella septempunctata* (M. P. Hassell, unpublished data). (c) Percentage of searching time of the ichneumonid wasp, *Diadromus pulchellus,* at different densities of leek moth pupae, *Acrolepia assectella,* per unit area (Noyes, 1974). (d) Distribution of *Syrphus* eggs in relation to the number of tests of the psyllid, *Cardiaspina albitextura,* per leaf surface (Clarke, 1963). (e) Number of parasitoid hours spent on plants of different *Pieris brassicae* density by the braconid, *Apanteles glomeratus* (Hubbard, 1977). (f) Percentage of searching time by the ichneumonid, *Nemeritis canescens,* at different densities of its flour moth host, *Ephestia cautella,* per container (Hassell, 1971a).

Chapter 5 then pursues the same theme somewhat further. There I discuss the phenomenon of mutual interference between searching predators and conclude with the interrelationships between this and predator aggregation.

51

AGGREGATIVE BEHAVIOR

Central to our discussion is the notion of a prey "patch." The term "patch" has been variously interpreted (Wiens, 1976), but we shall just view it as a spatial unit of the predator's foraging area—one whose appropriate dimensions are set not by what *we* perceive, but by the predator's foraging behavior itself. We should look to specific changes occurring in predator behavior that are associated with the recognition of patch and non-patch areas (Waage, 1977), just as in a more general context such behavioral changes help to distinguish between "trivial" movements within a habitat and migration between habitats (Kennedy, 1961, 1975; Southwood, 1962, 1977a, b).

There are various kinds of behavior that will result in predators tending to aggregate in patches of high prey density. To classify these, it is convenient to divide prey-finding into two distinct phases: the location of the prey patch—often the food-plant upon which the prey are feeding—followed by search for prey within the patch. The location of prey patches will usually depend upon visual or olfactory cues which may or may not be related to the density of prey within the patch. If they are so related, aggregation will be at least in part due to relatively long-range attraction of the predators. Two examples serve to illustrate this. Camors and Payne (1972) found that the parasitoid *Heydenia unica* responds directly to a volatile terpene released as a result of feeding by its bark beetle host; and Arthur (1966) showed the ichneumonid, *Itoplectis conquisitor,* to be attracted more to the red shoots of *Pinus sylvestris* than to the green ones, the reddening being due to infestations of its host, the moth *Rhyacionia buoliana.*

Sometimes the attraction is directly to the host individuals, even over considerable distances. Mitchel and Mau (1971) and Sternlicht (1973), for instance, provide clear examples of parasitoids that are attracted by their host's sex pheromones. More recently, examples of acoustical orientation to hosts have emerged. Cade (1975) has shown that the tachinid, *Euphasiopteryx ochracea,* that parasitizes the cricket, *Gryllus integer,* is attracted by the male cricket's song; and Soper, Shewell, and Tyrrell (1976) describe a sarcophagid parasitoid, *Colcondamyia auditrix,* which also responds to the mating song of its male hosts, in this case cicadas. In both instances, parasitized males were incapable of producing songs, an interesting means of preventing superparasitism.

Most examples of predator aggregative responses, however, will result primarily from patch specific behavior leading to longer periods of time being spent in those patches where the rate of encounter with prey is highest. The arrestant stimuli responsible for this are various and have been thoroughly discussed by Waage (1977). There may, for example, be a threshold rate of prey encounters, probably influenced by hunger level, below which the predator leaves the patch. Thus, Lindley (1974) found that immature lyniphiid spiders tend to disperse on gossamer threads from an area where no prey had been recently captured. Similarly, Turnbull (1964) showed that web-spinning spiders abandoned sites that were unprofitable, which led to spiders accumulating in regions of higher prey density. A different mechanism, shown by several predators, involves a change in searching pattern following feeding. This is well shown in Figure 4.2a and b for a "pseudo-predator", a housefly, searching for sugar droplets distributed in clumps on the floor of an arena (Murdie and Hassell, 1973). There is a pronounced increase in turning rate (klinokinesis) and a reduction in speed of movement (orthokinesis) when search is resumed after each feed. This behavior then decays to the prefeeding pattern within approximately 30 seconds if not reinforced by the

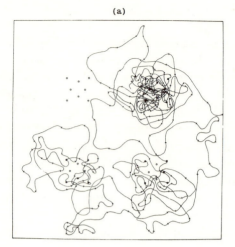

(a)

(b)

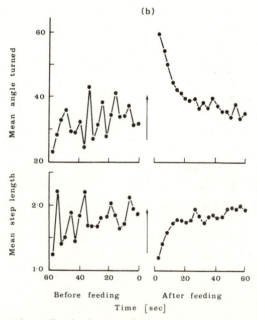

FIGURE 4.2. (a) Track of a housefly, *Musca domestica,* in an arena with four clumps of sugar droplets. Solid circles show where feeding occurred. The track is marked at 3-second intervals. (b) Mean angled turned (deg) and step length (cm) per second by houseflies in the 1-minute periods before and after feeding. Data pooled from 29 experiments (from Murdie and Hassell, 1973).

location of further droplets. A similar sequence was observed by Evans (1973) for the anthocorid predator *Anthocoris confusus* feeding on aphids, but here the turning behavior lasted for approximately 8 minutes, a difference probably related to the predator's search rate a' and the average patch size of its prey. Other examples of increased turning behavior are readily available, particularly from insects and mites (e.g. Laing, 1937; Fleschner, 1950; Banks, 1957; Bänsch, 1966; Chandler, 1969; Hassell and May, 1973).

Yet a third, distinct mechanism for predator aggregation is seen in some less mobile insect predators, such as some aphidophagous syrphid larvae, whose distribution depends largely upon the behavior of the adult female in distributing her eggs. An example of this is shown in Figure 4.1d and parallels the situation in insect parasitoids where the distribution of parasitoid progeny rests solely on the adult female's searching behavior.

The examples in Figure 4.1 do not immediately suggest a general form for a predator's response to prey distribution. Two are convex, two are concave, and two appear sigmoid. Indeed, a unique form of response is unlikely; it will depend in good measure on the many behavioral factors determining how long a predator is to remain in a patch. In this context, Waage (1977) lists four different mechanisms leading to quite different aggregative responses (Figure 4.3).

(a) *Fixed Number* mechanisms. The predator leaves a patch after a fixed number of prey have been captured. This is the "Hunting by Expectation" hypothesis of Gibb (1962).

(b) *Fixed Time* mechanisms. The predator leaves after a fixed amount of time has been spent in a patch. This is the "Hunting by Time Expectation" hypothesis of Krebs (1973).

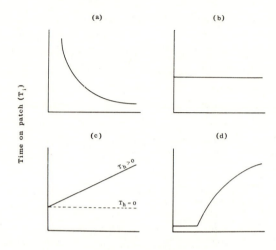

Prey density per patch (N_i)

FIGURE 4.3. Some hypothetical relationships between the time spent by a predator in a patch and the prey density per patch. (a) A fixed number of prey are eaten in a patch before leaving. (b) A fixed amount of time is spent in a patch before leaving. (c) A fixed amount of *searching* time is spent before leaving. This differs from (b) only if there is a finite handling time considered. (d) The predator leaves a patch when the rate of prey capture falls below a fixed level (after Waage, 1977).

(c) *Fixed Searching Time* mechanisms. The predator leaves after a constant *searching* time per patch.

(d) *Fixed Rate* mechanisms. The predator leaves when the rate of prey capture falls below a fixed threshold rate.

The fixed rate mechanism has recently been modeled by Murdoch and Oaten (1975). They made the following basic assumptions.

(1) A predator searches randomly while within a patch.

(2) A predator leaves a patch if no prey are found within a fixed threshold time θ. If a prey *is* caught before this time elapses, the "threshold clock" is reset to zero after the prey has been handled, and search resumes.

(3) There is a finite transit time for travel from patch to patch.

(4) The prey per patch are distributed according either to the Poisson or to the negative binomial distribution.

Murdoch and Oaten conclude that searching time per patch in relation to prey density tends to be sigmoid and that the transit time contributes significantly to the stabilizing effect of the aggregative response.

A somewhat similar, albeit more naive model, was outlined by Hassell and May (1974). They retain the threshold time θ but neglect transit time and do not specify any prey distribution. Thus the time spent per patch T_i is governed solely by the density of prey in that patch N_i and also by the search rate a' and the handling time T_h. Employing equation (3.1), the time required to encounter the first prey in a patch will be

$$T_1 = \frac{1}{a'N_i}. \tag{4.1}$$

If $T_1 \geq \theta$, the predator will have departed without encountering any prey; otherwise a prey is encountered and the predator remains in the patch searching for further prey. Similarly, the time taken to encounter the nth prey (from leaving the $n-1$th prey) becomes

$$T_n = \frac{1}{a'[N_i - (n-1)]}. \tag{4.2}$$

We now assume that the condition for leaving a patch ($T_n \geq \theta$) occurs after $n-1$ prey have been eaten, where

$$n - 1 = N_i - \frac{1}{a'\theta}. \tag{4.3}$$

To obtain the total time in the patch, we can make use of equation (3.6) which gives the number of prey eaten in a fixed period of time assuming random exploitation. Thus, identifying the $n - 1$ of equation (4.3) with the N_a of equation (3.6), we have

$$N_i - \frac{1}{a'\theta} = N_i \left[1 - \exp\left\{-a'\left(T_i - \left[N_i - \frac{1}{a'\theta}\right]T_h\right)\right\}\right].$$
(4.5)

to which we must add the constraint that $T_i = \theta$ if $T_1 \geq \theta$. This equation generates aggregative responses of the form of Figure 4.3d, although judicious use of the parameters θ, a' and T_h can make them appear more or less concave or convex. Waage (1977) has obtained similar curves from a much more complex behavioral model for aggregation by the parasitoid *Nemeritis canescens* in patches of high host density. Patch time is here determined primarily by two factors acting together. Firstly, the amount of patch odor (due to hosts) sets a level of responsiveness which decays with time on the patch. Secondly, any oviposition serves to increase this responsiveness by a set amount and so prolongs time on the patch. The parasitoid finally leaves the patch when this responsiveness decays below a threshold level.

From Figure 4.3, it seems clear that patch-time mechanisms determined by the rate of prey capture (i.e. Figure 4.3d) are likely to be more efficient in terms of the prey encountered per predator than fixed time mechanisms, and certainly more so than fixed number mechanisms. This question of the efficiency of patch-time allocation has prompted the development of several optimal foraging models (MacArthur and Pianka, 1966; Royama, 1971a, b; Emlen, 1973; Charnov, 1976; Cook and Hubbard, 1977), all of which attempt to compute the ideal allocation of foraging time to maximize prey capture rate and hence fitness. The tendency is always towards the reduction of all patches to the same rate of prey capture. In general, these theories are directed to vertebrates and especially birds, but that of Cook and Hubbard stands apart in being aimed at insect parasitoids. Their model is based directly on the random parasitoid equation (3.5), with the addition of a fixed total transit time between patches. The predictions are in the form of an optimal distribution of time among the optimal set of host patches to be visited. The crucial parameters that influence the time budget between patches are therefore the transit time and those of the random predator equation (a', T_h, T).

AGGREGATION AND STABILITY

The aggregation of predators where prey are abundant provides a potentially powerful stabilizing mechanism for predator-prey interactions, a conclusion that is supported by a variety of theoretical models (e.g. Hassell and May, 1973, 1974; Murdoch and Oaten, 1975; Murdoch, 1977). In effect, predator aggregation provides a partial refuge for the prey in low density patches, a situation that is not dissimilar from having complete prey refuges either in space or time (see below). An experimental system that points well to the importance of such spatial heterogeneity is due to Huffaker (1958) and Huffaker, Shea, and Herman (1969) working with two species of mite: a predatory mite, *Typhlodromus occidentalis,* and its orange-feeding prey, *Eotetranychus sexmaculatus.* At first, Huffaker employed a relatively simple system in which a variable number of oranges were dispersed among rubber balls on a tray. Each orange was partly covered with paraffin wax to present a constant area upon which the prey could feed. Initially, the prey were introduced on their own to illustrate the single species dynamics—a rapid increase of the population followed by a tendency toward limit cycles (Figure 4.4a). The introduction of predators to this system led to a single predator-prey oscillation, usually followed by the virtual extinction

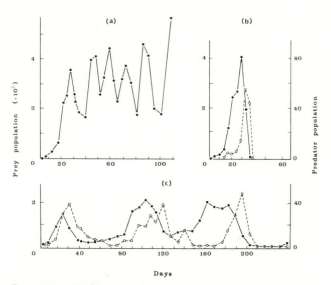

FIGURE 4.4. Predator-prey interactions between the mite, *Eotetranychus sexmaculatus* (●) and its predator, *Typhlodromus occidentalis* (○). (a) Population fluctuations of *Eotetranychus* without its predator. (b) A single oscillation of predator and prey in a simple system. (c) Sustained oscillations in a more complex system (after Huffaker, 1958), and figure taken from Hassell, 1976).

of both populations (Figure 4.4b). The complexity of the system was then increased by using 120 oranges, each with only one-twentieth of its surface exposed. The rapid dispersal of mites from orange to orange was impeded by placing partial barriers of vaseline between oranges. At the same time, dispersal was facilitated by placing small upright sticks at intervals throughout the system from which mites could launch themselves on silken strands carried by air currents. The results (e.g. Figure 4.4c) provide some of the finest examples of sustained predator-prey oscillations in a laboratory system. It is clear that the inclusion of considerable spatial heterogeneity, and the dispersal of predator and prey that goes with it, conferred marked stability on the populations.

60

A number of theoretical models support this conclusion. Vandermeer (1973), for example, showed that dispersal between patches and local population extinction can stabilize an otherwise unstable predator-prey interaction. Similarly, Maynard Smith (1974), Roff (1974), Hilborn (1975), and Hastings (1977) all discuss models in which there are a number of patches where predator and prey interact and among which they disperse. They too conclude that dispersal can contribute markedly to stability. In a rather different vein, Gurney and Nisbet (1978) have shown that the existence of a stable equilibrium with superimposed random elements of extinction and colonization of patches can account for the cyclic fluctuations observed in Huffaker's experiments.

A Simple Model

We now turn to a very simple model of predator aggregation, explored by Hassell and May (1973), that will further serve to indicate some of the features of prey and predator distributions that enhance stability. Our starting point is the Nicholson-Bailey model (2.5), but the total prey and predator populations in each generation are now distributed among n available patches. These patches may be leaves, individual plants, groups of plants etc., depending upon the range of movement and behavior of an individual predator. In each ith patch, we let the fraction of the prey population be α_i and the corresponding predator fraction be β_i, so that the sum of the α_i and of the β_i-values are each equal to unity. The function for predation in equation (1.1) now becomes

$$f(N_t, P_t) = \sum_{i=1}^{n} [\alpha_i \exp(-a\beta_i P_t)], \qquad (4.6)$$

which serves to distribute in each generation the N_t prey and P_t predators between the n patches in the proportions specified by α_i and β_i. Clearly, in the limiting case where the predators are equally distributed over all patches ($\beta_i = 1/n$), the system effectively collapses back to the Nicholson-Bailey model with random search.

The stability analysis of model (1.1) incorporating (4.6) above is given in Hassell and May (1973) and shows that, unlike the Nicholson-Bailey model, stability may now occur if there is a sufficiently uneven prey distribution and enough predator aggregation in patches of high prey density. The precise conditions for stability are

$$\lambda \sum_{i=1}^{n} [\alpha_i(a\beta_iP^*) \exp(-a\beta_iP^*)] < \frac{\lambda - 1}{\lambda}, \qquad (4.7)$$

indicating that stability hinges on the prey rate of increase λ, $\{\alpha_i\}$ and $\{\beta_i\}$.

Equation (4.6) provides a completely general model with which to explore the effects of different prey and predator distributions. It requires, however, separate calculation for the values of each special $\{\alpha_i\}$ and $\{\beta_i\}$ set that are used. As a result, general conclusions are more easily obtained by fixing upon one kind of prey distribution and a rather naive model for predator aggregation. Both will be made more realistic later in this chapter. We assume the prey distribution to have a single patch of high prey density containing a fraction α of the total prey, and to have the remaining $n - 1$ patches all with $(1 - \alpha)/(n - 1)$ prey. The predator distribution is defined by introducing a further parameter—the "aggregation index" μ:

$$\beta_i = c\alpha_i^{\mu}, \qquad (4.8)$$

where c is a normalization constant. Thus, the value of μ governs the degree of aggregation in high prey density patches as shown in Figure 4.5, with $\mu = 0$ corresponding to random search, and $\mu = \infty$ to all predators being in the

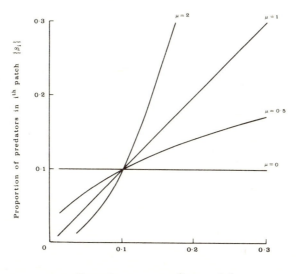

FIGURE 4.5. Some relationships from equation (4.8) between the proportion of searching predators $\{\beta_i\}$ and the proportion of prey $\{\alpha_i\}$ per patch using the different values of the aggregation index μ as shown (from Hassell and May, 1973).

single patch of highest prey density with other patches now being complete refuges from predation.

Using this particular model, Hassell and May showed there to be four parameters affecting stability, illustrated here from the stability boundaries in Figures 4.6a and b.

(1) μ, the predator aggregation index. Increasing values of μ tend to increase stability. Thus, predator aggregation alone can sometimes stabilize a model which otherwise is quite unstable.

(2) λ, the prey rate of increase. In all cases, stability breaks down abruptly as λ increases.

(3) α, the proportion of the prey in the high density patch. Stability is possible over a widening range of λ-values as α increases, although large values of $\alpha (> 0.5)$ no longer

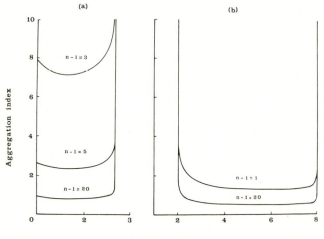

FIGURE 4.6. Stability boundaries between the aggregation index μ and the prey rate of increase λ from equation (4.6), where β_i is obtained from equation (4.8). The boundaries are shown for different values of (n − 1), the number of "low density patches," and for two values of α, the proportion of prey in the single "high density patch." (a) $\alpha = 0.3$; (b) $\alpha = 0.7$ (after Hassell and May, 1973).

permit stability at very low rates of prey increase. In addition, if α is small, and hence there is less contrast between high and low density patches, the minimum value of μ for stability increases. Indeed, if there is too little contrast in the prey densities per patch, no amount of predator aggregation may be sufficient for stability.

(4) (n − 1), the number of low prey density patches. For a given value of α, less predator aggregation is needed for stability as more low density patches are available. This is consistent with the conclusion of Maynard Smith (1974), Roff (1974), and Hilborn (1975) that stability is enhanced by increasing the number of spatial subunits.

Similar conclusions were reached by Bailey, Nicholson, and Williams (1962) who derived a criterion similar to (4.7) for the stability of a host-parasitoid system in which some hosts are more difficult to find than others. Clearly, the precise stability conditions will always be influenced considerably by the details of the prey and predator distributions. The tenor of the four conditions above, however, should still apply. We may be confident that predator aggregation will, in general, promote stability where prey are unevenly distributed, and that less aggregation will be required for more clumped prey distributions.

Refuges

A special case of non-random search, akin to the predator aggregation so far considered, is where some prey are completely free from predation within a temporal or spatial refuge. Spatial refuges can take a variety of forms lying between two extremes: (1) where a constant *proportion* of the prey population are protected, and (2) where a constant *number* are protected. Of these, constant proportion refuges appear to be the most common. For example, flour moth caterpillars (*Ephestia* spp.) are protected from being parasitized by the ichneumonid, *Nemeritis canescens,* when they are sufficiently deep in the flour medium to be out of reach of the parasitoid's ovipositor. In this way, only a *proportion* of the host's habitat is searched. Often, of course, the refuges cannot be so precisely defined. This is the case for the tightly-packed colonies of cabbage aphids *(Brevicoryne brassicae).* Those at the periphery are more susceptible to parasitism by the braconid, *Diaeretiella rapae,* than those towards the center, a proportionate effect that will increase with the area of the aphid colony (Akinlosotu, 1973). A similar observation has been made by Callan (1944) for a scelionid parasitizing the egg masses of the cacao stink-bug.

Temporal refuges, where prey and predators do not completely overlap in time, may also tend to protect a constant proportion of the prey. But here again, we can expect considerable variation, because the degree of asynchrony depends in part on a variety of physical factors. The way that age structure can also lead to an effective temporal refuge if some age classes of prey are not predated is discussed by Smith and Mead (1974) and May (1975a).

To explore the effect of refuges, we commence with a constant proportion of prey protected within the basic Nicholson-Bailey model (2.5). In each generation, a fraction γ of the prey are available to the predators, and hence a fraction $(1 - \gamma)$ within a refuge. This leads to the model

$$N_{t+1} = \lambda(1 - \gamma)N_t + \lambda\gamma N_t \exp(-aP_t)$$
$$P_{t+1} = \gamma N_t[1 - \exp(-aP_t)].$$

(4.9)

In the context of the "aggregation" function (4.6), this represents the case where $n = 2$, $\alpha_1 = \gamma$, $\alpha_2 = (1 - \gamma)$, $\beta_1 = 1$, and $\beta_2 = 0$. The detailed stability analysis is presented in Hassell and May (1973) and leads to the stability boundaries in Figure 4.7a. Clearly, proportionate refuges can lead to stability, but large areas of parameter space remain where there is either an unstable equilibrium (too few prey within the refuge) or no equilibrium at all (too many prey within the refuge).

Turning now to the constant number refuge, where N_0 prey are always protected, we have

$$N_{t+1} = \lambda N_0 + \lambda(N_t - N_0) \exp(-aP_t)$$
$$P_{t+1} = (N_t - N_0)[1 - \exp(-aP_t)],$$

(4.10)

which, in terms of equation (4.6), implies that $n = 2$, $\alpha_1 = (N_t - N_0)/N_t$, $\alpha_2 = N_0/N_t$, $\beta_1 = 1$ and $\beta_2 = 0$. This model includes the density dependent effect of having an increasing proportion of prey exposed to predation as

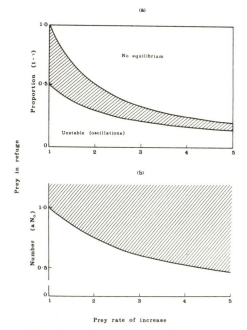

FIGURE 4.7. The effects of refuges on stability. (a) A constant proportion refuge. Stability boundaries from equation (4.9) between the proportion of prey in a refuge $(1 - \gamma)$ and the prey rate of increase λ. The hatched area indicates conditions for stability. (b) A constant number refuge. Stability boundaries from equation (4.10) between the number of prey in a refuge (scaled as aN_0) and λ. Stability is now enhanced since the region of "no equilibrium" has disappeared (after Hassell and May, 1973).

prey density increases. It should, therefore, be more stable than model (4.10), which is indeed apparent from Figure 4.7b where the region of "no equilibrium" has disappeared.

The qualitative effects of refuges can depend sensitively on the particular predator-prey model employed. This has led to conflicting statements in the literature. Maynard Smith (1974), for example, used the neutrally stable Lotka-Volterra model to show that only constant number refuges can contribute to stability. He showed that with a fraction

γ of the prey available at any time, then

$$\frac{dN}{dt} = N(r - \gamma bP)$$

$$\frac{dP}{dt} = P(-d + \gamma cN),$$

(4.11)

in which case the stability must remain unaltered since b and c are equally scaled. If, on the other hand, fixed number refuges are employed, then

$$\frac{dN}{dt} = rN - bP(N - N_0)$$

$$\frac{dP}{dt} = P[-d + c(N - N_0)],$$

(4.12)

and the density dependent effect so introduced inevitably converts the neutral stability to a stable equilibrium. Model (4.12) is therefore necessarily stable, in contrast to (4.10).

We may conclude, at least for systems with some time delays, that all refuges can contribute to population stability, but that this is most marked if a fixed number of prey tend to be protected. Stability from fixed proportion refuges depends upon a rather delicate balance between too few and too many prey protected.

More Realistic Distributions

Equation (4.6) permits a wide range of predator behavior, from random search ($\mu = 0$) to complete aggregation in the highest prey density patch ($\mu \to \infty$), and yet is sufficiently simple to respond well to an analytical treatment. Its foremost failings are that the prey distribution is excessively oversimplified and that the predators can respond only to the *proportion* of prey in each patch, and not to the *number* of prey present. Therefore, while pointing to some general features of spatial heterogeneity that are important, it still does not permit us to identify the critical components of more realistic predator responses. Ideally, we need to identify easily measurable parameters that characterize predator aggregative responses to realistic prey distributions and to show how these affect the dynamics of a predator-prey interaction. The model in this section, from Hassell and May (1974), is directed towards this goal, although simplifying assumptions continue to be necessary if the analytical nature of the model is to be retained.

Our starting point is given by equation (4.6) but now we replace the α_i and β_i term as follows

$$f(N_t, P_t) = \sum_{i=1}^{n} \left[\frac{N_i}{N_t} \exp \left(-a \frac{P_i}{P_t} P_t \right) \right], \qquad (4.13)$$

where N_i and P_i are the numbers of prey and predators in each ith patch. The "fractional predator" term, P_i/P_t, may now be identified with the total time spent in each patch, namely,

$$\frac{T_i}{T_0 + \sum_{i=1}^{n} T_i}, \qquad (4.14)$$

where T_i is the time spent per patch and T_0 is the "transit time" spent traveling between patches (Murdoch and Oaten, 1975). Equation (4.13) thus becomes

$$f(N_t, P_t) = \sum_{i=1}^{n} \left[\frac{N_i}{N_t} \exp \left(- \frac{a T_i P_t}{T_0 + \sum_{i=1}^{n} T_i} \right) \right], \qquad (4.15)$$

in which the original α_i and β_i terms have been dissected to reveal more meaningful parts. It may now be employed to explore the effect of particular prey and predator distributions.

In choosing a prey distribution—$p(j)$, the probability of finding j prey in a patch—we have again followed Murdoch and Oaten (1975) in using the negative binomial as a convenient and realistic clumped distribution for prey populations. It is one of a family of clumped distributions that have been very useful in describing natural populations, and is a clear improvement over the arbitrary assignments of fractions of prey per patch as discussed earlier. The details of the negative binomial have been well described in several texts (e.g. Pielou, 1969; Southwood, 1976) so that it suffices to mention here just that it is characterized by two parameters: the mean ($\bar{N} = N_t/n$) and a clumping index k. At the limits $k \to \infty$, $k = 1$ and $k \to 0$, a Poisson, a geometric, and a log series are obtained respectively. The way in which k expresses the clumping of the prey can be seen from the exact expression for the standard deviation relative to the mean in the negative binomial:

$$\frac{\sqrt{\text{variance}}}{\text{mean}} = \left[\frac{1}{k} + \frac{1}{\bar{N}}\right]^{1/2} . \qquad (4.16)$$

For an appropriate distribution of predator time—$T(j)$ the time spent in a patch with j prey—we have focused upon the generalized predator response in Figure 4.8. The significant components of this are the ratio of upper and lower asymptotes, defined by ϵ, and the transition J, shown here as a region, although in the analysis it is taken as the point at which the response begins to rise from the lower asymptote ϵT.

The way in which these distributions may be approximated and the detailed stability analysis that results are presented in Hassell and May (1974). The emphasis here is on the emergent conclusions and, in particular, on the features affecting stability, bearing in mind that equation (4.15) reverts to the unstable Nicholson-Bailey model (2.5) when T_i is constant for all n patches.

The existence of a stable equilibrium depends upon the quantities ζ, aJ, λ, k, and ϵ, where λ is the prey rate of

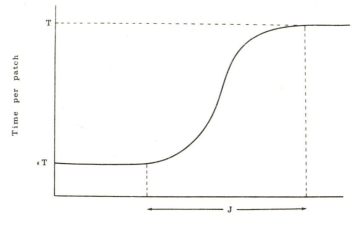

FIGURE 4.8. A schematic illustration of a general predator response to a patchily distributed prey. Stability is enhanced by a large difference between the minimum (ϵT) and maximum (T) time spent per patch, and by the prey equilibrium falling within the transition region J (from Hassell and May, 1974).

increase, k is the exponent of the negative binomial distribution, and ϵ is as defined above. The quantity ζ is a dimensionless combination introducing the transit time T_0:

$$\zeta = \frac{1}{k} \left[\frac{T_0}{nT} \right] , \qquad (4.17)$$

where T is defined in Figure 4.8, and the quantity aJ is the combination of searching efficiency and the transition point, also shown in Figure 4.8. Since the equilibrium prey population N^* is of the order of $1/a$ (especially as $\lambda \to 1$), this combination indicates the magnitude of N^* relative to J, of critical importance since stability is possible only if N^* falls in the region where T_i is an increasing function of prey density per patch. We should expect, therefore, a "window" of aJ-values, commencing at $aJ = 1$ (i.e. $N^* \simeq J$), in which a stable equilibrium is possible.

71

Four conclusions stand out from the stability analysis.

(1) The relative difference between the maximum (T) and the minimum (ϵT) times spent per patch (i.e. $1/\epsilon$) affects the width of the "aJ-window" and hence the region of stability, as shown in Figure 4.9. Increasing values of ϵ reduce the stability domains up to a critical value of ϵ when a stable equilibrium becomes impossible. The predators are now spending too great a fraction of their time in the low prey density region $(j < J)$.

(2) Stability is increased as the prey become more clumped (i.e. as k of the negative binomial decreases). This is shown from curves A and B in Figure 4.9.

(3) Stability is also increased as more time is spent in transit between patches (T_0), just as found by Murdoch and

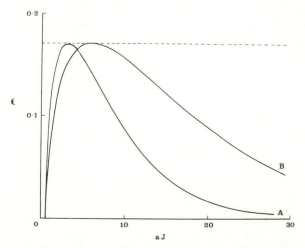

FIGURE 4.9. Stability boundaries between the ratio of the minimum to maximum time spent per patch, ϵ, and the combination aJ discussed in the text. The boundaries are shown for two levels of prey clumping. Curve A: $\zeta = 0$, $k = 0.4$. Curve B: $\zeta = 0$, $k = 0.2$. Stability is increased by a reduction in ϵ (and hence a larger difference between the minimum and maximum time spent per patch) and by the prey population becoming more clumped (after Hassell and May, 1974).

Oaten (1975) and Murdoch (1977). It arises simply from more time being spent moving between patches when average prey densities are low rather than high. The result is a density dependent effect whereby the net searching efficiency increases as average prey density increases.

(4) As in all such predator-prey models, stability decreases as the prey rate of increase λ is increased.

We may conclude, therefore, that stability is markedly affected by the transit time, the prey distribution, and by two critical components of the predator's response to this distribution: the ratio of times spent in patches of low and high prey density and the range of prey densities over which T_i increases.

A likely complication in the real world is that predators will rarely forage over a single type of patch. A hierarchy of relevant patch sizes is much more likely. For instance, there could be one pattern of distribution from plant to plant and another from leaf to leaf within a plant. If this were the case, equation (4.6) would become

$$f(N_t, P_t) = \sum_{i=1}^{n} \alpha_i \sum_{j=1}^{x} \alpha_j \exp(-a\, \beta_i \beta_j P_t), \qquad (4.18)$$

where there are n plants with a fraction α_i prey and β_i predators per plant and x leaves per plant with the corresponding fractions being α_j and β_j. Several qualitative features emerge without exploring the model in detail (R. M. May, pers. comm.).

(1) In general, such "double aggregation" has the effect of compounding two patterns of prey and predator clumping and so tends to enhance stability.

(2) Inclusion of a transit time T_0 will shift the emphasis more toward the plant-to-plant dispersion since T_0 between plants will be much longer than between leaves.

(3) Random search at both levels leads to recovery of the Nicholson-Bailey model.

(4) Random search over just one of the levels leads back to equation (4.6) (unless, as is likely, the prey-per-leaf set $\{\alpha_j\}$ is itself a function of $\{\alpha_i\}$, whereupon a more complex model results for aggregation at the leaf level).

A Detail Independent Model

The aggregation models so far described have been made progressively more complicated in an effort to capture some essential components of predator and prey distributions. Unfortunately, such complexity leads rapidly to analytically intractable models which are a deterrent against further elaboration of the systems. A way around this is to seek a *minimally* complicated means of introducing non-random predator search, while at the same time retaining the essential dynamical effects of predator aggregation. An elegant example of this is due to May (1978) who follows Griffiths and Holling (1969) in adopting the negative binomial to describe the distribution of parasitoid encounters with hosts. This leads to the population model

$$
\begin{aligned}
N_{t+1} &= \lambda N_t \left[1 + \frac{aP_t}{k} \right]^{-k} \\
P_{t+1} &= N_t \left[1 - \left(1 + \frac{aP_t}{k} \right)^{-k} \right],
\end{aligned}
\tag{4.19}
$$

where the parameter k—the exponent of the negative binomial—captures the degree of predator aggregation, albeit in a most detail independent way. We are thus not concerned here with any assumptions about the prey's distribution, nor with any details of predator behavior that lead to aggregation in patches of high prey density. We are instead looking at the end product of this behavior, manifest as a clumped distribution of predator attacks. Predator aggregation is strongest when $k \to 0$ and weakest as

$k \to \infty$, when the distribution of predator attacks becomes Poisson and the Nicholson-Bailey model is reobtained.

There remains however, as May (1978) has shown, a specific biological meaning that can be attached to k; one that ties this model closely to the more detailed ones already described. Let us consider a system in which the searching predators are distributed over n patches according to some *arbitrary* distribution in which the mean number of predators per patch is P and the variance in the numbers of predators per patch is σ^2. The coefficient of variance of the number of predators per patch, CV_P, is thus given by

$$(CV_P)^2 = \sigma^2/P^2. \tag{4.20}$$

We will assume that *within* a patch the predator attacks are distributed randomly according to a Poisson series, so that the attacks upon *all prey in all patches* are now distributed according to a joint distribution that depends on the details of the arbitrarily chosen predator distribution, convolved with a Poisson within each patch. Let us now suppose that we approximate this combined distribution by a negative binomial with the same mean and same variance as the precise combined distribution. (The negative binomial is exactly appropriate here if the chosen predator distribution is a Pearson Mark III). We now find, as expected, that the negative binomial has a mean aP and, more interestingly, that

$$k = \frac{1}{(CV_P)^2}. \tag{4.21}$$

The formal proof of this is given in May (1978). Again we find that the Nicholson-Bailey model is recovered if the predators are exactly uniform in their distribution over patches, since $CV_p = 0$ and hence $k \to \infty$. The important conclusion, therefore, is that k from equation (4.19) may be directly interpreted in terms of the coefficient of variance of the numbers of predators per patch, and we may retain our conception of patches and roving predators when dealing with this model.

The provenance of such heuristic models as (4.19) lies in parasitology, in particularly with Crofton (1971a, b), in whose population model the parasites are overdispersed among their hosts in accord with the negative binomial (see May, 1977a, for a re-examination of this work). The evidence supporting the negative binomial in this parasitological context is considerable and has recently been reviewed by Anderson (1978) and Anderson and May (1978). In addition, the distribution gains support from a variety of models for the infection of hosts by parasites, all of which lead to a negative binomial distribution of parasites per host (Bradley and May, 1978).

Equation (4.19) therefore provides us with a three-parameter model representing the outcome of predator aggregative behavior, the dynamical properties of which have been fully discussed by May (1978). By setting $N_{t+1} = N_t = N^*$ and $P_{t+1} = P_t = P^*$ in the usual way, the equilibrium populations are given by

$$N^* = \frac{\lambda}{\lambda - 1} P^*$$

$$P^* = \frac{k}{a} (\lambda^{1/k} - 1). \tag{4.22}$$

For these to be stable requires that

$$\frac{\lambda - 1}{\lambda} > k \left[\frac{\lambda^{1/k} - 1}{\lambda^{1/k}} \right] \tag{4.23}$$

which is true if, and only if, $k < 1$. Thus for $k > 1$, there are always expanding oscillations; for $k < 1$, the system is always stable, at first oscillatorily and then as k becomes even smaller, exponentially.

76

Very small values of k (i.e. strong predator aggregation) have yet an additional effect: the term $k\lambda^{1/k}$ in equation (4.23) becomes exceedingly large and hence gives unreasonably high values of N^* and P^*. (Exactly the same is true as μ from equation (4.8) increases.) It therefore becomes particularly appropriate to set some upper limit upon the prey population by including a density dependent rate of increase as described in Chapter 2 using the Nicholson-Bailey model. To do this, we adopt the function (2.10b), which gives the model

$$N_{t+1} = \lambda N_t e^{-gN_t} \left[1 + \frac{aP_t}{k} \right]^{-k}$$
$$P_{t+1} = N_t \left[1 - \left(1 + \frac{aP_t}{k} \right)^{-k} \right].$$

(4.24)

The stability properties of this model are best displayed by following Beddington, Free, and Lawton (1975, 1976) in defining an additional parameter q, the extent to which the predators depress the prey equilibrium below the "carrying capacity" (see page 24) (i.e. $q = N^*/K$). Recalling the definitions introduced in equations (2.10a) and (2.10b), we have

$$K = (\log_e \lambda)/g,$$

(4.25)

and thence

$$q = gN^*/\log_e \lambda.$$

(4.26)

Thus, $q = 0$ corresponds to the limit where there is no prey density dependence ($g = 0$) and $q = 1$ to the absence of predators. This definition enables stability diagrams to be drawn in terms of q against λ for various levels of "predator aggregation" k, as shown in Figure 4.10. From 4.10a, we see that in the limit $k \to \infty$, the Nicholson-Bailey model with prey density dependence is recovered and hence also the stability diagram (2.7). As the clumping of predator attacks increases ($k \to 0$), there is at first little change

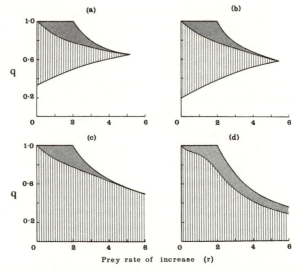

FIGURE 4.10. This figure illustrates the stability boundaries for the system described by equation (4.24). The depression of the prey equilibrium, q, is shown as a function of the prey rate of increase (r or $\log_e \lambda$), shown for four values of k (the exponent of the negative binomial distribution). The stippled and hatched areas denote the regions of exponential and oscillatory damping respectively. (a) $k = \infty$. (b) $k = 2$. (c) $k = 1$. (d) $k = 0.1$.

in the boundaries, but as soon as $k \leqslant 1$, condition (4.23) is satisfied and the model becomes widely stable.

It is likely that models such as these, in which particular distributions of prey and searching predators need not be specified, will have an increasingly important part to play in exploring the effects of non-random search in more complex predator-prey systems. Obvious candidates for this are the multi-prey and multi-predator systems described in Chapters 7 and 8.

SUMMARY

Non-random search, where more time is spent in patches of high prey density, is likely to be quite general among predators and parasitoids and can have a pronounced effect on the stability of a population interaction.

This is first demonstrated from a model in which the prey distribution is arbitrarily fixed and the predators' response to the prey's distribution is described by a simple "aggregative index." Stability is then found to depend in large part on the degree of clumping of the prey and on the strength of the predator aggregative response. Prey refuges are also considered as a special case of this model where there are but two prey patches, in only one of which the predator can search. Two types of refuge are identified: those where a constant proportion of prey are protected from generation to generation and those where a constant number are so protected. Both are potential stabilizing mechanisms, with the "constant number" refuge being the more significant.

A more detailed model, where the negative binomial describes the prey's distribution, and the predators' response to this is more in accord with observed behavior, supports the general findings of the simpler model. Stability is enhanced by increased prey clumping, more marked predator aggregation and, in this case, also by more time spent in transit between patches as first found by Murdoch and Oaten (1975).

These models with explicit prey and predator distributions soon become complex and difficult to explore analytically. Alternative models that capture the spirit of predator aggregation in a detail independent way are available and should prove invaluable in facing up to more complex problems. One of these, based on a negative binomial, rather than Poisson, distribution of encounters with prey is outlined and shown to be stable for all values of k (the exponent of the negative binomial) less than unity.

CHAPTER FIVE

Mutual Interference

As predators aggregate in patches of high prey density, it is increasingly likely that they will encounter each other while searching for prey, which in turn may lead to an increased tendency toward dispersal. The effects of such encounters between searching predators are classified under the general heading of "mutual interference." Mutual interference and the aggregative behavior discussed in the previous chapter are, in reality, closely intertwined. For simplicity, however, we commence with interference that is divorced from any responses to prey distribution, for which most information is available.

Within the confines of a laboratory cage, some predators and parasitoids have been observed to react markedly to the presence of other searching individuals nearby. The parasitoids *Diaeretiella rapae* that parasitizes aphids and *Diadromus pulchellus* that parasitizes leek moth pupae, for example, react to encounters with other individuals by an increased tendency to fly off the host food-plant. This is evident from Figures 5.1a and b showing emigration rates to increase with parasitoid density. The same tendency has been observed in *Nemeritis canescens* (see below), both when adult females encounter each other and after the detection by a female of a host that has already been parasitized (Rogers, 1972). Similar behavior has been observed among arthropod predators. Kuchlein (1966) has shown that increasing densities of the predatory mite, *Typhlodromus longipilus,* leads to increased dispersal rates from experimental leaf discs containing prey mites, and Michelakis (1973) has observed the tendency of the larvae of *Coccinella septempunctata* to drop from plants following encounters.

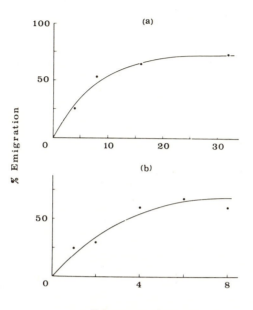

Parasitoid density

FIGURE 5.1. The effect of female parasitoid density on parasitoid emigration from an experimental cage. (a) The braconid, *Diaeretiella rapae* (Akinlosotu, 1973). (b) The ichneumonid, *Diadromus pulchellus* (Noyes, 1974). Both curves fitted by eye.

The common effect of such mutual interference is to reduce the available searching time in direct proportion to the frequency of encounters. We would thus expect to find that the searching efficiency per predator over the experimental period, namely,

$$a = a'T = \frac{1}{P_t} \log_e \left[\frac{N_t}{N_t - N_a} \right] \qquad (5.1)$$

(derived from equation (2.4)),

declines as predator density increases. That this is often the case, at least from laboratory experiments, is clear from Figure 5.2. Interference may also be manifest in other ways. For instance, predator females may lay fewer eggs as

81

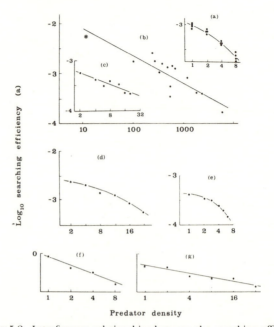

FIGURE 5.2. Interference relationships between the searching efficiency ($\log_{10} a$) and density of searching parasitoids or predators. (a) *Pseudeucoila bochei* parasitizing *Drosophila* larvae (Bakker, Bagchee, van Zwet, and Meelis, 1967). (b) *Nemeritis canescens* parasitizing larvae of *Anagasta kühniella* (Hassell and Huffaker, 1969). The ringed point is discussed in the text and comes from generation 15 of the room B system shown in Figure 5.7. (c) *Cryptus inornatus* parasitizing cocoons of *Loxostege sticticalis* (Ullyett, 1949a). (d) *Encarsia formosa* parasitizing the whitefly, *Trialeurodes vaporariorum* (Burnett, 1958a). (e) *Chelonus texanus* parasitizing eggs of *Anagasta kühniella* (Ullyett, 1949b). (f) *Coccinella septempunctata* feeding on *Brevicoryne brassicae* (Michelakis, 1973). (g) *Phytoseiulus persimilis* feeding on deutonymphs of *Tetranychus urticae* (Fernando, 1977). Further examples where similar relationships have been observed are listed in Table 5.1.

predator density increases, as demonstrated by Evans (1976) for *Anthocoris confusus* feeding on aphids (Figure 5.3a). Alternatively, crowding of parasitoids may lead to a pronounced shift towards males in the sex ratio of the eggs laid. This has been observed by Wylie (1965) for *Nasonia vitripennis* parasitizing housefly pupae (Figure 5.3b), by Viktorov (1968, 1971) for *Trissolcus grandis* and *T. volgensis* parasitizing the bug, *Eurygaster integriceps,* and by Viktorov and Kotshetova (1973) for *Dahlbominus fuscipennis* parasitizing cocoons of the sawfly, *Neodiprion sertifer.* These effects are primarily due to females laying fewer fertilized (= female) eggs when crowded and hence most likely to be found among Hymenopterous parasitoids in which haplodiploidy is the normal method of sex determination.

Results such as those in Figure 5.2 led Hassell and Varley (1969) to propose a simple modification of the Nicholson-

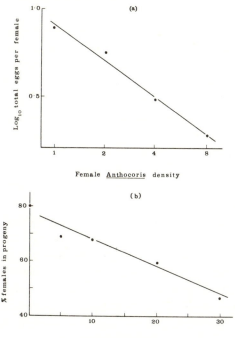

FIGURE 5.3. The effect of female density on (a) the fecundity of the predatory bug, *Anthocoris confusus* (Evans, 1973), and (b) the sex ratio of the progeny of the parasitoid, *Nasonia vitripennis* (Wylie, 1965).

Bailey model, based on the assumption that there is a linear relationship on logarithmic scales between searching efficiency a and predator density:

$$a = QP_t^{-m}, \tag{5.2}$$

where Q and m are constants.* Substituting for a in the Nicholson-Bailey model (now a special case when $m = 0$), gives the model

$$N_{t+1} = \lambda N_t \exp(-QP_t^{1-m})$$
$$P_{t+1} = N_t[1 - \exp(-QP_t^{1-m})], \tag{5.3}$$

a modification that has a marked affect on stability. Instead of always being unstable, there can now be a stable equilibrium given suitable values of the interference constant m and the prey rate of increase λ. The stability analysis leading to this conclusion is given in Hassell and May (1973), and the stability boundaries between m and λ are shown here in Figure 5.4. The third parameter in the model, Q, has no effect on stability but does, of course, affect the equilibrium levels of the populations.

While equation (5.2) provides a significant description of the data in Figure 5.2, it is clear that the relationships cannot really be linear over a wide range of predator densities. Royama (1971), Rogers and Hassell (1974), and Beddington (1975) have all stressed that searching efficiency cannot continue to rise indefinitely as predators become increasingly scarce. It is much more likely that the response is curvilinear, tending to level off at very low predator densities when interference is negligible. Indeed, this is apparent from some of the relationships in Figure 5.2.

---

* The confused dimensions of Q are rather unfortunate, but equation (5.2) has sufficient a pedigree that it is better retained to be consistent with past usage.

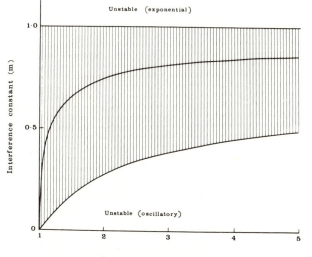

Prey rate of increase (λ)

FIGURE 5.4. Stability boundaries from equation (5.3) in terms of the interference constant m and the prey rate of increase λ. The hatched area denotes the conditions for stability and is divided into two regions; one of exponential damping lying above that of oscillatory damping. The line between these regions indicates the conditions for the most rapid approach to the equilibrium (from Hassell and May, 1973).

A MORE BEHAVIORAL MODEL

Steps toward a less empirical description of interference relationships have been taken by Rogers and Hassell (1974) and Beddington (1975). Spatial heterogeneity continues to be excluded, but now the predators are assumed to encounter each other at random, defined by a rate b (analogous to the rate of encounters with hosts a'). The authors further assume that following each encounter there is a period of time "wasted," T_w, during which there is no search. The only point of difference between the models lies in whether a predator can interfere with others during T_w. Beddington assumes that they can, while Rogers and

Hassell's predators leave the prey area during T_w and hence are unavailable for further interference. The difference may be an important one biologically but has little effect on the outcome of the models. Beddington's model, however, is the more tractable mathematically and thus the one to be outlined here.

We begin with the simple functional response model from Chapter 2, namely,

$$N_e/P_t = a'T_sN_t, \tag{5.4}$$

where N_e is the number of encounters with prey and T_s is the time spent searching, and observe that T_s now depends not only on the total handling time per predator, but also on the total time "wasted" due to interference:

$$T_s = T - \left[T_h \frac{N_e}{P_t} + T_s bT_w(P_t - 1) \right], \tag{5.5}$$

where T is the total time initially available. Neglecting handling time for convenience (which makes little differ-ence to the outcome of the analysis), we have

$$T_s = \frac{T}{1 + bT_w(P_t - 1)}, \tag{5.6}$$

which on substitution into equation (5.4) gives

$$\frac{N_e}{P_t} = \frac{a'N_tT}{1 + bT_w(P_t - 1)}. \tag{5.7}$$

This equation has an intuitive derivation similar to that of the disc equation (3.4); the only difference is that the dis-count term is now $bT_w(P_t - 1)$ for interference rather than $a'T_hN_t$ for handling time.

As with the disc equation, equation (5.7) is an instan-taneous expression that does not cater for prey exploita-tion. To overcome this, we integrate over the duration T as outlined in Appendix I to derive the more usable expression,

$$N_a = N_t \left[1 - \exp\left(-\frac{a'TP_t}{1 + bT_w(P_t - 1)} \right) \right], \quad (5.8)$$

one that reduces again to the Nicholson-Bailey equation once $bT_w = 0$. The kinds of interference relationship generated by this model are shown in Figure 5.5. They are always curvilinear and, as pointed out by Free, Beddington, and Lawton (1977), have an asymptotic slope of 1.0 at high predator densities. Altering the value of $a'T$ merely shifts the relationship "up and down," while changing bT_w moves it from "side to side."

In short, the inclusion of explicit terms for the rate of encounters among predators and for the time "wasted" at each encounter leads to an effective value of m, the interference constant, lying somewhere between 0 and 1. The results in Figure 5.2 can therefore all be interpreted in

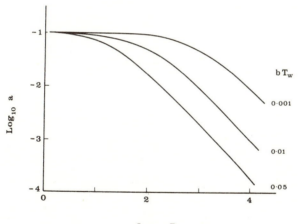

FIGURE 5.5. Interference relationships obtained from equation (5.8) for three values of bT_w, given that $a'T = 0.1$. All the curves have an asymptotic slope of 1.0 at high predator densities (after Free, Beddington, and Lawton, 1977).

terms of this simple behavioral model, those fitted by straight lines representing linear approximations to segments of the total, curvilinear response.

An advantage of the stability boundaries in Figure 5.4 is that they remain appropriate, regardless of the interference sub-model adopted. The only requirement is that the value of m be taken as the slope of the response evaluated at the equilibrium predator density P^*. Thus for any relationship of the type in Figure 5.5 or 5.6, the procedure for determining stability is as follows.

(1) Calculate the potential equilibrium density of predators P^* using the particular model adopted. This is the point where $\lambda f(N_t, P_t) = 1$.

(2) Evaluate the slope of the curve at this equilibrium point ($m = m^*$ in Figure 5.6).

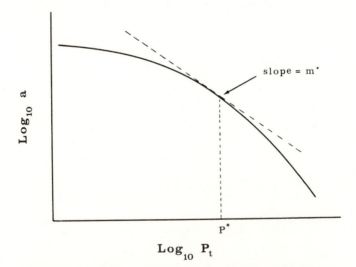

FIGURE 5.6. A generalized curvilinear interference relationship between the searching efficiency a and the predator density, on logarithmic scales. m^* is identified as the slope of the curve evaluated at the equilibrium predator population P^* (from Hassell and May, 1973).

(3) The model will be stable if

$$1 > m^* > 1 - \frac{\lambda - 1}{\lambda \log_e \lambda},$$ (5.9)

and Figure 5.4 may be used by substituting m^* for m. A fuller account of this recipe is given in Hassell and May (1973).

VALUES OF m FROM LABORATORY AND FIELD

There is no doubt that interference can be most important under laboratory conditions. But is it an important part of natural predator-prey interactions, or largely a laboratory artifact as suggested by Griffiths and Holling (1969)? That it is *never* significant is unlikely. Why else should so many parasitoids have evolved threat display and aggressive behavior to other females in the vicinity? Scelionids (egg parasitoids), for example, often show aggressive behavior towards other females on an egg mass, as observed by Hidaka (1958), Subba Rao and Chacko (1961), Wilson (1961), Hokyo and Kiritani (1963, 1966), and Safavi (1968). Similarly, Spradbery (1970) found that female *Rhyssa persuasoria*, an ichneumonid parasitoid of wood wasp larvae, will threaten and, if need be, fiercely drive an "intruding" female from the same area of tree trunk, and may maintain that "territory" for several days. Interestingly, he has also shown (Spradbery, 1969) that the ovipositing *Rhyssa* ignores the very close presence of its cleptoparasitoid, *Pseudorhyssa sternata*, indicating that the aggression is not directed to *any* parasitoid in the vicinity.

To pursue this further, an obvious step is to compare the values of m obtained in the laboratory and in the field, as done here in Table 5.1. It is striking from these results that the field examples tend to provide the highest m-values. These, however, must be treated with caution since in obtaining m, predator density appears in both dependent and independent variables; any errors in estimating P_t then bias the relationships to give an artificially high value of m (Hassell and Varley, 1969). Unfortunately, in all but one of the field examples in Table 5.1, we find that the apparent interference could be due solely to sampling errors. Only *Alaptus fusculus,* a mymarid parasitoid attacking *Mesopsocus* eggs, shows an interference relationship that is statistically significant (Broadhead and Cheke, 1975).

Recently Free, Beddington, and Lawton (1977) have argued on theoretical grounds that we are unlikely to find significant amounts of interference at *equilibrium population densities.* They take equation (5.8) which, within our general model (1.1), yields the function

$$f(N_t, P_t) = \exp\left[-\frac{a'TP_t}{1 + bT_w(P - 1)}\right] \qquad (5.10)$$

and then following Hassell and May's (1973) recipe, find the value of $m*$ to be given by

$$m* = \frac{bT_w}{a'T} \log_e \lambda. \qquad (5.11)$$

Arguing that the attack rate a' and encounter rate b will take roughly similar values since both are dependent on predator activity, and that $\log_e \lambda$ will usually be close to unity, it becomes clear that $m*$ will scale approximately as T_w/T. This implies that $m*$, at least near the equilibrium and in a homogeneous environment, is largely determined by the ratio of the time wasted by a predator following a single interference encounter to the duration of its searching lifetime, a ratio that must often be considerably less than the m-values shown in Table 5.1. (This conclusion is along much the same lines as that mentioned in Chapter 3: the effect of handling time on stability depends on the ratio T_h/T).

TABLE 5.1. Some values for the interference constant m obtained from laboratory and field studies.

Species	m	Field (F) or Lab (L)	Para- sitoid or Predator	Author(s)
Phytoseiulus persimilis	$\begin{cases}0.18* \\ 0.44**\end{cases}$	L	Predator	Fernando (1977)
Anagyrus pseudococci	0.18	L	Parasitoid	M. J. Berlinger (pers. comm.)
Dahlbominus fuscipennis	0.28	L	Parasitoid	Burnett (1956)
Aphytis coheni	0.33	L	Parasitoid	D. J. Rogers (pers. comm.)
Leptomastix flavus	0.33	L	Parasitoid	M. J. Berlinger (pers. comm.)
Aphidius uzbeckistanicus	0.35	L	Parasitoid	Dransfield (1975)
Coccinella septempunctata	0.38	L	Predator	Michelakis (1973)
Cryptus inornatus	0.38	L	Parasitoid	Ullyett (1949b)
Encarsia formosa	0.38	L	Parasitoid	Burnett (1958)
Alaptus fusculus	0.39	F	Parasitoid	Broadhead & Cheke (1975)
Bracon hebetor	0.44	L	Parasitoid	Benson (1973)
Bathyplectis anurus	0.47	L	Parasitoid	Latheef, Yeargen & Pass (1977)
Telenomus nakagawai	0.48	F	Parasitoid	Nakasuji, Hokyo & Kiritani (1966)
Cyzenis albicans	0.52	F	Parasitoid	Hassell & Varley (1969)
Chelonus texanus	0.54	L	Parasitoid	Ullyett (1949a)
Aptesis abdominator	0.60	L	Parasitoid	von B. Sechser (pers. comm.)
Diaeretiella rapae	0.65	L	Parasitoid	Chua (1975)
Nemeritis canescens	0.67	L	Parasitoid	Hassell (1971a)
Pseudeucoila bochei	0.68	L	Parasitoid	Bakker, Bagchee, van Zwet & Meelis (1967)
Cratichneumon culex	0.86	F	Parasitoid	Hassell & Varley (1969)
Olesicampe benefactor	0.91	F	Parasitoid	Ives (1976)
Apanteles fumiferanae	0.96	F	Parasitoid	Miller (1959)
Phygadeuon dumetorum	1.13	F	Parasitoid	Kowalski (1977)

* Adult females searching for deutonymph prey.
** Adult females searching for larval prey.

Other evidence that interference is negligible under equilibrium conditions is hard to find. Free, Beddington, and Lawton quote as an example the study of Hassell and Huffaker (1969) who analyzed the interaction of the parasitoid *Nemeritis canescens* and its stored product host *Anagasta kühniella* in two room ecosystems. In room A, host and parasitoid interacted for 23 generations with the parasitoids showing only a seven-fold fluctuation in densities as shown in Figure 5.7a. Under these conditions, close to an apparent equilibrium, no correlation was evident between $\log a$ and $\log P_t$. The room B interaction presented a

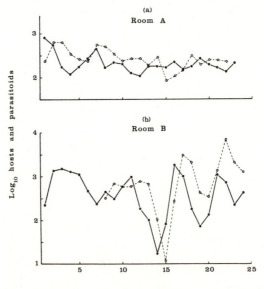

FIGURE 5.7. Population fluctuations of the adult flour moth, *Anagasta kühniella* (●), and its larval parasitoid, *Nemeritis canescens* (○), in two room systems. Room A: food supply for *Anagasta* larvae was confined to 340 containers with 1.2 g food per container. Room B: as Room A, but *Nemeritis* absent for the first 7 generations. Food supply to the hosts was then increased to 7.2 g per container (after Hassell and Huffaker, 1969).

marked contrast. It differed in the hosts having a larger food supply and the parasitoids being only introduced in the eighth generation. Six hundred-fold fluctuations in parasitoid density were now observed (Figure 5.7b) and a significant interference relationship obtained (see Figure 5.2b). Free, Beddington, and Lawton suggest that this relationship is given by the excessively high parasitoid densities found in room B. But this is not entirely the case. Also very important to the relationship is the circled point in Figure 5.2b, where the *lowest* parasitoid density recorded (generation 15 in Figure 5.7b) has given the *highest* searching efficiency.

This example, therefore, does not fully support the notion that there is insignificant interference at equilibrium levels, presumably because the data is roughly linear over a very wide range of parasitoid densities. Under natural conditions, free from the influence of a relatively small confined system, curvilinear relationships more of the form in Figures 5.5 and 5.6 are to be expected. Only relatively high predator densities can now lead to an interference relationship and hence Free, Beddington, and Lawton's conclusion will be much more likely to apply.

This argument in no way denies the importance of interference when predator densities are well above their equilibria. This may be of especial importance, for example, following the successful introduction of a parasitoid (or more occasionally a predator) for the biological control of a pest. Faced with a superabundance of hosts, the parasitoids increase rapidly to exceptional densities, before causing a decline in host numbers and hence subsequently in their own. An observed feature of some of these introductions is the accelerating rate of spread of the parasitoids as they become increasingly abundant. A well-documented instance of this is the spread of *Olesicampe benefactor*, a parasitoid of the larch sawfly in Canada, shown here in

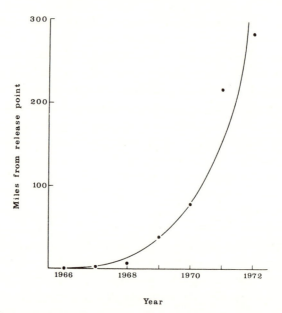

FIGURE 5.8. The spread from its release point of the ichneumonid parasitoid, *Olesicampe benefactor*. The parasitoid was first released in 1961 in an attempt to control the larch sawfly, *Pristiphora erichsonii* (information in *Forestry Report* (Canadian Forestry Service, Edmonton, Alberta), August 1973, 3, 1–2).

Figure 5.8. Since an exponential population increase in conjunction with random dispersal would tend to give a linear relationship between distance and time, it is likely that interference is here serving to increase the rate of dispersal, just as observed on a minute scale for *D. rapae* and *D. pulchellus* in Figure 5.1. Similar very high rates of dispersal have been mentioned by Townes (1971) for other parasitoids following introductions for biological control.

AGGREGATION AND INTERFERENCE

So far we have considered interference divorced from any predator response to a patchily distributed prey. In this event, interference would seem a maladapted behavior since its only effect would be to reduce the overall prey consumption or number of hosts parasitized. But once it is acknowledged that prey are heterogeneously distributed, and in particular that they occur in discrete units of the habitat, the situation is profoundly altered. Interference leading to dispersal can now serve to distribute predators more efficiently among all patches, and so lead to a better foraging strategy. We turn now, therefore, to interrelationships between aggregation and interference and find that they are more subtle than might be expected.

In the first place, we note that within a confined system, aggregation in patches of high prey density can lead to more marked interference, simply by increasing the frequency of encounters between the searching predators. This is certainly the case for *Nemeritis canescens* parasitizing the larvae of the stored product moth, *Ephestia cautella* (Hassell, 1971a, b; Hassell and Rogers, 1972). In Figure 5.9a and b we see contrasted the interference relationships from two similar cage systems. In one, 564 *Ephestia* larvae were distributed over the entire floor area of the cage (50 cm × 50 cm) in a thin layer of wheat toppings. The resulting relationship (Figure 5.9a) is curvilinear, of the form predicted by Rogers and Hassell (1974) and Beddington (1975), and shows little sign of interference except at the highest parasitoid densities. In the other system, the same cages were employed, but the 564 hosts were now confined within 16 small containers as follows:

2 of 128; 2 of 64; 3 of 32; 3 of 16; 3 of 8; 3 of 4.

The difference is conspicuous. There is now a pronounced linear relationship (Figure 5.9b), with $m = 0.69$, resulting from the strong tendency for *Nemeritis* to aggregate on the few high host density containers, and hence to increase the rate of mutual encounters.

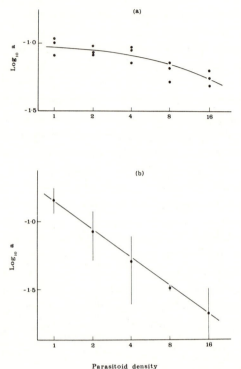

FIGURE 5.9. Interference relationships for *Nemeritis canescens* searching for larvae of the flour moth, *Ephestia cautella*. (a) Results of a 24-hour experiment in which 564 *Ephestia* were contained in a thin layer of wheat toppings on the floor of a cage of area 50 cm × 50 cm. The curve shown was fitted by eye. (b) As (a) but the 564 hosts were now confined in 16 small containers as described in the text. Means and 95% confidence limits are shown, and the regression line, $y = 0.85 - 0.69 x$ ($p < 0.001$).

We must beware, however, of ascribing such relationships entirely to the behavioral effects of mutual encounters. Free, Beddington, and Lawton (1977) have shown that marked aggregation can of itself lead to *apparent* interference, even if true interference is lacking (i.e. $bT_w = 0$). This they call "pseudo-interference." They illustrate the phenomenon by adopting equation (4.8) to describe the predator distribution within the general model (4.6). The number of prey eaten in the ith patch is therefore given by

$$N_{ai} = \alpha_i N_t [1 - \exp(-a'_p T \alpha_i{}^\mu P_t)], \qquad (5.12)$$

where N_t is the total number of prey in all patches and a'_p is the search rate per patch (handling time is again omitted, since it has little bearing on the argument).

An arbitrary, clumped prey distribution $\{\alpha_i\}$ is now selected, and values for a'_p and the aggregation index μ assigned. By varying the predator density and finding in each case the total number of prey eaten from all patches, it is possible to calculate the *overall* search rate for each predator density. This is done by taking the n patches present as a single homogeneous area and then applying equation (5.1), namely,

$$a' = \frac{1}{P_t T} \log_e \left[\frac{N_t}{N_t - N_a} \right], \qquad (5.13)$$

where a' is now the overall search rate, N_a is the total number of prey eaten in all patches, and T the duration of the interaction. In this way, pseudo-interference can be demonstrated by comparing a' with P_t, as done in Figure 5.10a. A fuller picture is given in Figure 5.10b. Here the amount of pseudo-interference is shown to increase as predator aggregation increases, an effect which becomes more noticeable as the prey becomes more clumped. In addition, m^* can be shown to increase with increasing numbers of low density patches or with a larger fraction of the prey in the highest density patch. These results are in close qualitative agreement with those of Hassell and May (1973), as indeed they should be if m^* fully encapsulates the stabilizing effects of the more complex model.

The explanation that underlies this rests solely in the differential exploitation of patches of different prey den-

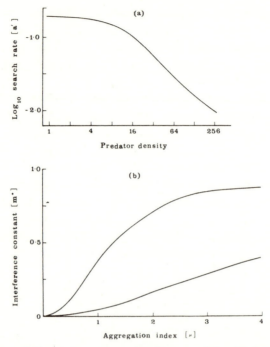

FIGURE 5.10. (a) An apparent interference relationship between the search rate a' and predator density, arising solely from predator aggregation in patches of high prey density. The search rate a' over all patches was calculated from equation (5.13) applied to equation (5.12), where $a'_p = 0.25$, $T = 1$, $N_t = 100$, $\mu = 3$ and $\{\alpha_i\} = 0.8$; 0.2. (b) The relationship between the pseudo-interference constant m^* evaluated at the predator equilibrium P^* and the degree of predator aggregation μ. The apparent interference is entirely due to the non-random search of the predators among prey patches, and becomes more marked with increasing predator aggregation. The two curves illustrate how pseudo-interference is enhanced by a more clumped prey distribution. Parameter values: $a'_p = 0.15$, $T = 1$, $N_t = 100$, $\{\alpha_i\}$ upper curve = 0.8; 0.2, and $\{\alpha_i\}$ lower curve = 0.6; 0.4.

sity. As Free, Beddington, and Lawton state: "Parasites aggregate in [high density] regions because they are initially the most profitable. This behavior enables a higher proportion of the hosts in the whole area to be parasitized than would be possible with random search." This is illustrated in Figure 5.11, showing how aggregation leads to higher overall search rates than would random search, except when predator density is high and exploitation therefore considerable. The "profitability" of continued aggregation now falls rapidly and a more even or random search strategy becomes preferable. The stabilizing effect of aggregation is thus due to the prey in the low density patches being somewhat protected and hence in "partial refuges." This is encapsulated by m^*.

Essentially the same phenomenon of pseudo-interference has been noticed in different contexts by Stinner and Lucas (1976), Cook and Hubbard (1977), and Münster-Swendsen and Nachman (1978). Yet a further demonstration of it is due to May (1978). He begins with

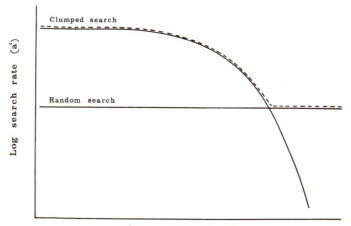

FIGURE 5.11. A schematic picture to illustrate both the increased searching efficiency at low and intermediate predator densities and the apparent interference relationship that arise from predators always aggregating in patches of high initial prey density rather than searching at random. The broken line illustrates a more prudent predator strategy in which aggregation gives way to random search at high predator densities.

equation (4.19), where the parameter k was used to capture the essence of predator aggregation. It is now a straightforward step to follow the method in Hassell and May (1973) and obtain an expression for the slope of the log a versus log P_t curve at the equilibrium predator density:

$$m^* = 1 - \left[\frac{k(1 - \lambda^{-1/k})}{\log_e \lambda} \right]. \tag{5.14}$$

This expression now permits us to take the stability boundaries in Figure 5.4 and on them to pinpoint the curves of m^* against λ for various values of k. The result is shown in Figure 5.12. Notice, as mentioned earlier, that the model is always stable if $k < 1$ and unstable for $k > 1$, and that for extremes of predator aggregation (e.g. $k = 0.1$) pseudo-

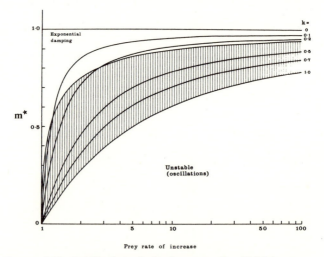

FIGURE 5.12. Stability boundaries from equation (5.14) in terms of the pseudo-interference constant m^* and the prey rate of increase λ. A locally stable equilibrium is only possible if k lies between 0 and 1. Other contours for k are also shown within the stable area, namely $k = 0.1$, 0.2, 0.5, and 0.7. The hatched area distinguishes the region of damped oscillations from that of exponential damping. Note that the outer boundaries are identical with those in Figure 5.4.

100

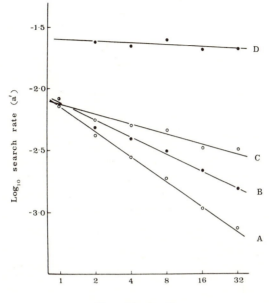

Nemeritis density

FIGURE 5.13. Relationships between the search rate a' and the density of searching *Nemeritis canescens*. The difference between slopes A and C is explained by behavioral interference and that between C and D by pseudo-interference. Further details are given in the text.

interference is such that damped oscillations give way to exponential damping.

An example that actually pinpoints the components of an interference relationship in the laboratory is shown in Figure 5.13. It is based on the experiments of Hassell (1971a, b), again using *Nemeritis canescens* parasitizing *Ephestia cautella*. The larvae were confined to 15 small containers in densities of 4 to 128 per container, and exposed to 1, 2, 4, 8, 16, or 32 parasitoids for 24 hours, at the end of which time parasitism was scored. Continuous observations were made during each experiment which yielded estimates for

101

(1) the total time per parasitoid spent on all containers,
(2) the time spent on containers of different host density, and
(3) the frequency of encounters on a container leading either to a parasitoid leaving the container or to an interruption of its probing behavior without leaving.

Lines A to D in Figure 5.13 are of progressively decreasing slope and are explained as follows.

Line A ($m = 0.67$). This is the all-inclusive interference relationship obtained from the results summed over all 15 patches. The search rate per hour is obtained from equation (5.13) where T is 24 hrs, N_t is the total host density of 532 per experiment, and N_a is the total number of hosts parasitized.

Line B ($m = 0.45$). This represents the relationship after allowance for the results in Figure 5.14, where each parasitoid is seen to spend less time on the host areas as parasitoid density increases. The value of T in equation (5.13) now varies from 21.6 hours for $P_t = 1$ (i.e. 2.4 hours spent off host containers) to 11.1 hours for $P_t = 32$.

Line C ($m = 0.27$). This now takes account of the observed number of encounters between parasitoids on the containers that leads to interruption of their probing for hosts (data given in Hassell and Rogers, 1972). It is assumed that probing is resumed after 1 minute, a guesstimate based on the continuous observations. Line C therefore represents the relationship after all observed *behavioral* interference has been abstracted. It suggests that pseudo-interference accounts for the remaining negative slope (0.27) and hence nearly a third of the overall interference relationship.

Line D ($m = 0.03$). This differs from line C only in that account is taken of the distribution of attacks and of the time spent in the n different patches. Thus, a' is to be

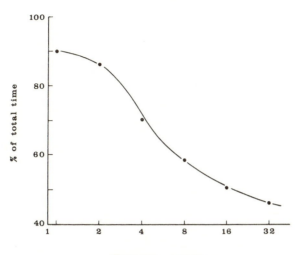

FIGURE 5.14. The relationship between the proportion of time spent by an individual *Nemeritis canescens* on host containers and the density of *Nemeritis* within the experimental cage.

calculated from

$$a_D' = \frac{1}{n} \sum_{i=1}^{n} \left[\frac{1}{P_t T_i} \log_e \left(\frac{N_i}{N_i - N_{ai}} \right) \right], \qquad (5.15)$$

where N_i, N_{ai}, and T_i are the number of hosts, number of hosts parasitized, and time spent respectively on the ith container. By taking full account of the actual searching time per container, shown in Figure 5.15, and the resulting number of hosts parasitized per container, a_D' is the best measure so far of the real, average search rate in a patchy environment. The increased level of line D merely reflects how the search rate is increased when changing from assumed random search (lines A to C) to the observed aggregative responses of *Nemeritis* from Figure 5.15.

The fact that line D is virtually horizontal is encouraging since it shows all interference to have been successfully

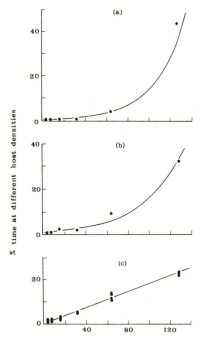

% time at different host densities

Host density per patch

FIGURE 5.15. Relationships showing the reduction in the aggregative response of an individual *Nemeritis canescens* as the number of searching parasitoids increases. (a) One parasitoid; $\log_{10} y = -0.505 + 0.016x$. (b) Two parasitoids; $\log_{10} y = 0.003 + 0.012x$. (c) Four, eight, sixteen and thirty-two parasitoids; $y = 0.413 + 0.18x$ (after Hassell, 1971a).

abstracted. Thus, of the original value of $m = 0.67$, 0.22 is due to behavioral interference causing *Nemeritis* to leave host patches and 0.18 is due to behavioral interference causing probing to be interrupted, leaving about 0.27 due to the differential exploitation of patches arising from non-random search (i.e. pseudo-interference).

This example further emphasizes that the stabilizing effects of both behavioral interference and predator aggregation can be encapsulated within the same parameter m.

The implication is that interference models such as (5.3) join ranks with equation (4.19) as the simplest ways of incorporating the effects of non-random search in predator-prey models.

Finally, we return to Figure 5.11 and make some concluding remarks on the interrelationship of aggregation and interference. Predators in general should spend more time searching in patches of high prey density than where prey are scarce, simply because the rate of prey encounters is likely to be greater than if search were random. Only at high, non-equilibrium predator densities is this not true. Under these conditions, the patches of abundant prey are becoming heavily exploited, and a rather more even search pattern should be adopted. This is indicated by the broken line in Figure 5.11. Behavioral interference is a means of promoting this transition, and does so by enhancing the dispersal of predators from patches in which the prey have already been, or are likely to become heavily exploited.

SUMMARY

Mutual interference among searching predators is a widespread phenomenon, at least under laboratory conditions, and a potentially powerful stabilizing mechanism in population interactions. It is unlikely to be as pronounced under natural conditions, although evidence from evolved aggressive behavior and dispersal rates indicates that it is of some significance. The interrelationship of mutual interference and aggregation is complex. Marked aggregation will increase the probability of mutual predator encounters which may in turn tend to increase their dispersal among patches. Aggregation alone can lead to apparent interference simply due to the uneven exploitation of patches. Evaluating this "pseudo-interference" is another means of encapsulating the stabilizing effects of non-random search.

CHAPTER SIX

The Predator Rate of Increase

All the predator-prey models so far discussed have assumed that each prey killed yields a constant number of predator progeny in the next generation. In this respect the models have been most appropriate to parasitoid-host interactions and less suitable for predators, whose rate of increase is unlikely to be such a simple linear function of the number of prey attacked. It is therefore important to discover how the dynamics of these predator-prey interactions differ from the parasitoid-host ones already described. To pursue this, we shall first review some relevant data, then seek a suitable form for the function describing the per capita predator rate of increase from equation (1.2), namely,

$$Q(N_t, P_t) = P_{t+1}/P_t, \qquad (6.1)$$

and finally make use of this to explore the dynamics of a simple predator-prey model.

THREE COMPONENTS

Under constant environmental conditions, the overall rate of increase of a predator population will depend largely on three components, each of which will be a function of the rate at which the average predator can locate and consume suitable prey. These components are (1) the developmental rate and (2) the survival rate of each larval instar, and (3) the fecundity of the adults. Our task, therefore, is to specify how each of these will change with the number of prey eaten. We have already seen that this depends on such factors as the prey density (Chapter 3), the predator density, and the relative distributions of predators and prey (Chapters 4 and 5). But, for simplicity, and also because of the availability of data, we will consider just a single predator searching for a homogeneous prey population. In this way, we will determine how the functional response alone can influence the three components above. A fuller treatment with a wider range of examples is given in Lawton, Hassell, and Beddington (1975) and Beddington, Hassell, and Lawton (1976).

Developmental Rates

On eclosion, a parasitoid larva is normally faced with abundant food, sufficient for all the larval instars. Only if there are excess larvae within the same host, or if for some reason the host is much smaller than average, will development tend to be slowed. The situation is very different, however, for true predators. Each larval stage must usually seek prey which are often much smaller than the predator, making several items necessary for continued development. Some of this food must inevitably be allocated to maintenance metabolism, so that growth will cease if consumption falls below a certain level. Above this threshold (which should become higher in successive instars as maintenance energy requirements increase), the energy allocated to growth will become a function of food intake. The simplest assumption, and still a reasonable one, is that this is a linear relationship within each instar such that

$$G = p(I - \alpha) \tag{6.2}$$

where G is the growth rate, conveniently measured as weight gain per unit time, I is the ingestion rate or biomass of prey consumed per unit time, α is the threshold ingestion rate below which growth ceases, and p is a constant.

Two examples of arthropod predators conforming to this model are given in Figure 6.1.

If we now let W be the total weight gain during an instar, then the ratio W/G will define the duration of the instar d, the inverse of which is the developmental rate. This argument ignores the fact that several arthropod predators are able, when food is scarce, to molt to the next instar at lower body weights than when food is abundant. Indeed, a few predators such as the spider *Linyphia triangularis* will molt in the complete absence of food, so that successive instars weigh progressively less, until death intervenes (Turnbull,

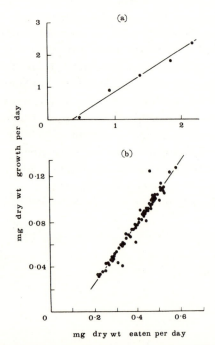

FIGURE 6.1. Predator growth rates as a function of their feeding rate. Both are linear relationships of the form specified by equation (6.2). (a) Final instars of the hemipteran, *Notonecta undulata* (Toth and Chew, 1972); $G = 0.99I - 0.46$. (b) Fourth instars of the spider, *Linyphia triangularis* (Turnbull, 1962); $G = 0.27I - 0.11$.

108

1962). If, however, we assume that W is indeed a constant, rather than a function of prey consumption, a convenient model for developmental rate becomes

$$\frac{1}{d} = \frac{P}{W} (I - \alpha). \tag{6.3}$$

Assuming now that the ingestion rate I is proportional to the number of prey eaten, namely,

$$I = k(N_a/P_t) \tag{6.4}$$

where k is a constant depending on the biomass of each prey eaten, the developmental rate can be related to prey density N_t using a convenient functional response model such as the disc equation (3.4). We now obtain the expression

$$\frac{1}{d} = \frac{P}{W} \left[\frac{ka'TN_t}{1 + a'T_hN_t} - \alpha \right], \tag{6.5}$$

which predicts a relationship with prey density that follows a negatively accelerating rise to a plateau and has a positive intercept on the X-axis, as shown by the examples in Figure 6.2. Other examples are provided by Rivard (1962), Dixon (1970), Fox (1973), Glen (1973), and Hodek (1973).

Survival Rates

We now turn briefly to the relationships between predator survival rates and prey density which, together with those defining developmental rates, are an essential ingredient in determining how prey density influences adult recruitment and hence the predator's overall rate of increase.

Let us consider the survival rate over a time period of one larval instar as a function of the average prey density available during that period. Unless each predator is genetically identical, starvation should occur at some characteristic mean ingestion rate μ_I with the population as a whole showing variation about this mean value. If we as-

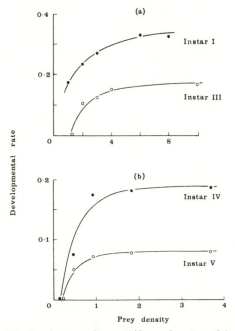

FIGURE 6.2. Developmental rates ($1/d$) as a function of the average prey density available during the developmental period. The curves are fitted by eye and are of the general form of equation (6.5). (a) Instars I and III of the coccinellid, *Adalia decempunctata* (Dixon, 1959). (b) Instars IV and V of the hemipteran, *Notonecta undulata* (Toth and Chew, 1972).

sume that the proportion of the population starving is normally distributed about the mean, with standard deviation σ_I, the proportion surviving to complete development S, at an ingestion rate I, will be given by

$$S = \frac{1}{\sqrt{2\pi}} \int_{-\infty}^{z} \exp\left(-\frac{z^2}{2}\right) dz, \tag{6.6}$$

where

$$Z = \frac{I - \mu_I}{\sigma_I}.$$

This treatment is reminiscent of, but less detailed than, that of de Jong (1976) who developed a model of feeding and survival among arthropod competitors.

Figure 6.3c shows some predicted relationships between survival S and prey density N_t. They were obtained by combining the relationship between S and I in Figure 6.3a (from equation (6.6)) with the functional responses in Figure 6.3b (from equation (3.4)), using equation (6.4) to relate the two. It is encouraging that these hypothetical relationships are of the same general form as found from experiments on a variety of arthropod predators. Two such examples are given in Figure 6.4 and others in Dixon (1959, 1970), Toth and Chew (1972), and Glen (1973).

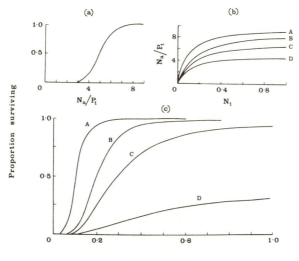

FIGURE 6.3. Hypothetical relationships between (a) the proportion of individual predators surviving to the end of an instar and their mean feeding rate during that instar, and (b) predator feeding rates and prey density. The former is defined by equation (6.6) and the latter by the disc equation (3.4) for four values of a' and T_h. Combining (a) with curves A to D in (b) produces the four curves in (c), which relate survival to prey density. These hypothetical curves are to be compared with the experimental ones in Figure 6.4 (from Beddington, Hassell and Lawton, 1976).

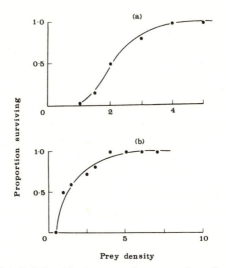

FIGURE 6.4. Relationships between the proportion of predators surviving to the end of a particular instar and the mean prey density during that instar. (a) First instar of the coccinellid, *Adalia bipunctata* (Wratten, 1973). (b) Survival through the first and second instars of the spider, *Linyphia triangularis* (Turnbull, 1962). Both curves have been fitted by eye.

In passing, we should note that, although treated separately, both the survival and developmental rates as influenced by prey consumption are likely to be interrelated, with survival tending to fall as the duration of an instar is prolonged. Several experimental examples of this are reviewed by Beddington, Hassell, and Lawton (1976).

Adult Fecundity

Parasitoids exhibit a wide range of feeding habits as adult females. Some emerge with sufficient protein, lipid, and carbohydrate reserves from the larval stage to maturate their full complement of eggs, while others must feed to do so. They may feed on host haemolymph or on food that is quite independent of their hosts, such as pollen,

112

nectar, and aphid honeydew. Despite this range of behavior, it is probably true that the fecundity of a female parasitoid is limited largely by the number of hosts that she can locate and rather little by nutritional considerations. It is this assumption that makes models of the form of equation (1.1) so appropriate to parasitoids.

Predatory arthropods also show very diverse feeding habits as adults. Some species (e.g. some Drosophilidae, Phaoniinae, and Syrphidae) have predatory larvae but non-predatory adults which, like the comparable parasitoids, are dependent on pollen and perhaps other food sources for full egg maturation. The majority, however, are also predatory as adults so that fecundity now depends considerably on prey consumption during the adult stage.

Sometimes the larval nutritional history may also be important in that it can affect adult size and hence the egg complement. But in this section we shall disregard this and assume that fecundity is solely determined by prey availability to the adults.

We commence by arguing, as for larval growth, that some of the food assimilated by the adult female predator must be allocated to maintenance metabolism, and hence be unavailable for egg maturation. There will thus be a threshold prey ingestion rate b below which reproduction ceases, but above which there is some dependence between fecundity F and ingestion rate I. By again assuming a linear relationship, we have

$$F = y(I - b), \tag{6.7}$$

where y is a constant. This model is well supported by several experimental examples such as those displayed in Figure 6.5.

By once more utilizing equation (6.4) and the disc equation (3.4), an explicit expression for fecundity in terms of prey density is obtained,

$$F = y \left[\frac{ka'TN_t}{1 + a'T_hN_t} - b \right], \tag{6.8}$$

that predicts fecundity to rise at a decreasing rate towards an upper asymptote as prey density increases. This, too, is supported by a wide range of examples (e.g. Figure 6.6), similar in form to the typical type II functional response but tending to be displaced along the prey axis until prey density is sufficient to achieve the threshold ingestion rate b.

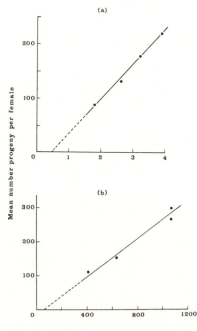

FIGURE 6.5. Reproductive rate as a function of the feeding rate per female predator. Both lines have been drawn by eye and are of the form of equation (6.7). (a) The waterflea, *Daphnia pulex* var. *pulicaria* (Richman, 1958). (b) The hemipteran, *Podisus maculiventris* (Mukerji and Le Roux, 1969).

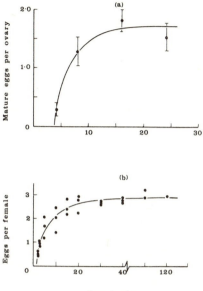

Prey density

FIGURE 6.6. Predator fecundity as a function of prey density. Both curves are fitted by eye and are of the form of equation (6.8). (a) The coccinellid, *Adalia decempunctata* with 95% confidence limits shown (Dixon, 1959). (b) The mite, *Phytoseiulus persimilis* (Pruszyński, 1973).

A PREDATOR-PREY MODEL

Ideally, a general model for the predator rate of increase should describe how development rates, survival, and fecundity change with prey consumption through a series of instars. The inevitable result would be a model of daunting complexity where the functional, interference, and aggregative responses all combine to affect each component. Variable development rates are especially awkward since they would force us to abandon our time-independent difference equations in favor of something mathematically more difficult.

Beddington, Free, and Lawton (1976) bypass these problems by making some effective simplifications, while still retaining a crucial feature of predator reproduction; namely, that some prey must be caught and consumed before any reproduction can occur. They commence with an equation of the form of (6.7), but now relate the adult fecundity F to the number of prey eaten *throughout the predator's life:*

$$F = c \left[\frac{N_a}{P_t} - \beta \right], \tag{6.9}$$

where c and β are constants displayed in Figure 6.7. Equation (6.1) may now be expressed as

$$Q(N_t, P_t) = \frac{P_{t+1}}{P_t} = c \left[\frac{N_a}{P_t} - \beta \right] \quad \text{if } \frac{N_a}{P_t} > \beta \tag{6.10a}$$

$$Q(N_t, P_t) = \frac{P_{t+1}}{P_t} = 0 \qquad\qquad \text{if } \frac{N_a}{P_t} \le \beta. \tag{6.10b}$$

Thus, c is the efficiency with which consumed prey are converted into new predators and β is the minimum prey consumption for any reproduction to occur.

They now assume that N_a is defined by the Nicholson-Bailey equation (2.4) (rather than the instantaneous disc equation as done for equation (6.8)), which on substituting into (6.10a) gives

$$\frac{P_{t+1}}{P_t} = c \left[\left\{ \frac{N_t}{P_t} [1 - \exp(-aP_t)] \right\} - \beta \right]. \tag{6.11}$$

Finally, they include a prey population with a density dependent rate of increase to provide their completed model:

$$N_{t+1} = N_t \exp[r(1 - N_t/K) - a\,P_t] \tag{6.12}$$

$$P_{t+1} = c[\{N_t[1 - \exp(-a\,P_t)]\} - \beta\,P_t].$$

This, therefore, differs from model (2.12) only in having a

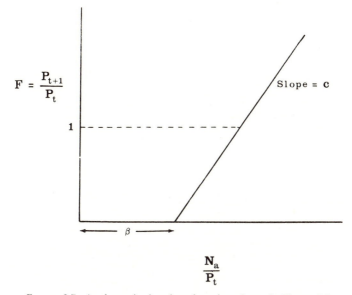

FIGURE 6.7. A schematic view, based on data shown in Figure 6.5, of the relationship between the per capita predator rate of increase and the number of prey eaten throughout the predator's life. The constant β represents the minimum prey consumption for any reproduction to occur, and the slope c expresses the efficiency of converting prey eaten into predator offspring. The broken line at $F = 1$ denotes the predator equilibrium P^* where $P_{t+1} = P_t$ (after Beddington, Free, and Lawton, 1976).

threshold prey consumption before reproduction can occur (i.e. $\beta > 0$), and in permitting $c \neq 1$.

In this way a model emerges in which predator reproduction hinges upon prey density. The model also includes the realistic feature of a density dependent prey growth rate but, on the other hand, lacks realism in the predators having a constant search rate a (i.e. no handling time, interference, or aggregative responses). All the same, it will serve to highlight a fundamental difference between the stability properties of predator-prey models where $\beta > 0$ and those of parasitoid-host models in which $\beta = 0$.

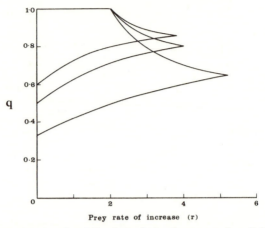

q

Prey rate of increase (r)

FIGURE 6.8. Local stability boundaries from equation (6.12) in terms of q, the depression of the prey equilibrium, and $r(= \log_e \lambda)$, the predator rate of increase. The boundaries are shown for three values of $c\beta$ ($c\beta = 2$, 1 and 0 in order of increasing stable space), and those for $c\beta = 0$ are identical to those in Figures 2.7 and 4.10a (after Beddington, Free, and Lawton, 1976).

We begin by viewing in Figure 6.8 some of the local stability boundaries for equation (6.12) for various values of β (or more strictly, for values of the combination $c\beta$, upon which the stability boundaries actually depend). When $\beta = 0$, the situation describes a parasitoid-host interaction and the boundaries are identical to those in Figure 2.7. The effect of increasing β, and so becoming more like a true predator, is to reduce this area of stable parameter space. The boundaries could, of course, be considerably enlarged by introducing some stabilizing predator behavior such as interference or aggregation, but this would not alter the qualitative effect of a reduction in stable space as β increases.

Increasing β has a further important effect. When $\beta = 0$, the equilibrium values, N^* and P^*, are globally stable, but this ceases to be true whenever $c\beta > 0$. We now have a situation in which local and global properties are different:

for each equilibrium point there is a domain of local stability (Figure 6.9), outside which cyclic or chaotic patterns of population behavior will occur. These stable domains are very sensitive to the value of $c\beta$, becoming extremely small for $c\beta > 4$ and, of course, immense as $c\beta \to 0$.

In simple terms, the parasitoid-host model, obtained when $c\beta = 0$, may be stable irrespective of the initial host and parasitoid population sizes. This, however, is not true

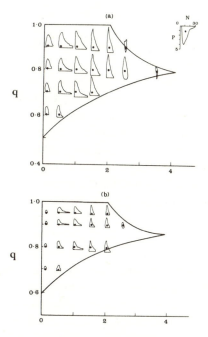

Prey rate of increase (r)

FIGURE 6.9. Local stability boundaries as in Figure 6.8, but now showing the domains of attraction for equilibria derived from a grid of values for q and r, assuming that the carrying capacity $K = 10$. The domains of each equilibrium (shown as a dot) are drawn to the same scale given in the inset. The equilibria are globally stable when $c\beta = 0$, but become locally stable as soon as $c\beta > 0$ and rapidly become very small as $c\beta$ increases. (a) $c\beta = 1$; (b) $c\beta = 2$ (after Beddington, Free, and Lawton, 1976).

119

of predators for whom the domains of local stability become smaller as the predators need to eat more prey before breeding (increasing β), and as they become more efficient at converting prey eaten into predator offspring (increasing c).

SUMMARY

A crucial difference between predators and parasitoids lies in the factors affecting their reproductive rates per generation. Unlike a parasitoid, a predator must locate and consume several prey items before developing to maturity. Any shortage of prey eaten will lead to a reduced developmental rate, survival rate, and adult fecundity. A full picture of predator reproduction should therefore include the influences of prey density, prey distribution, and predator density on the predator feeding rate and hence on development, survival, and fecundity.

Such a picture is not available at present but, as a first step, this chapter deals with the dependence of predator reproduction on prey density. This leads to a population model which departs from those of previous chapters in having a threshold prey consumption β below which there is no predator reproduction. The effect of this threshold is at least two-fold. It renders the interactions less stable than the equivalent parasitoid-host system and, perhaps more important, it introduces equilibria that are no longer globally stable. As the domains of local stability get smaller, the initial predator and prey population sizes become increasingly important. If these initial values are a long way from the equilibrium, the interaction may collapse rather than approach a stable point. This feature of the model could have important implications for the use of predators in biological control programs, suggesting that we should set a high premium on ensuring a suitable predator-prey ratio at the start of the program. This point is further discussed in Chapter 9.

The influence of prey density on the predator's rate of development has also been noted but its consequences neglected. To remedy this, we would have to abandon our simple time-independent difference equations in favor of mathematically less tractable models. This problem demands attention, since variable developmental rates are likely to have profound effects on the interacting populations.

CHAPTER SEVEN

Polyphagous Predators

The majority of the population models described in the previous chapters consider single predator-single prey species systems. This is the simplest situation and one that is appropriate to many questions on how predators can affect a prey's population dynamics. We shall now abandon this restraint and consider some ways in which polyphagous predators may affect the dynamics of their prey, an important step to take since relatively few predators and parasitoids are truly monophagous.

The central problem to be faced is how, if at all, polyphagous predators can affect coexistence in a system of competing prey species. Several classic studies have been addressed to this question (see Connell, 1975, for a review). In one of these, Paine (1966) found that, following the removal of the starfish *Pisaster ochraceus* from an area of seashore, competition among the prey (limpets, chitons, mussels, and barnacles) soon led to extinctions. After one year, six species had been replaced, with the mussel, *Mytilus,* and the goose barnacle, *Mitella,* finally dominating the system. Clearly, *Pisaster* as a "top predator" is playing a crucial role in maintaining the prey species diversity. But what are the essential ingredients for a predator to play this role? That it must be polyphagous is obvious. What is less clear is the relative importance of a high searching efficiency, of random or non-random search, of any preferences for certain prey species, and of the ability to "switch" from one prey species to another as their relative abundance changes. These problems are amenable to an analytical treatment, but first we must be clear on some aspects of preference and switching in polyphagous predators.

PREFERENCE

Preference for a particular prey is normally measured in terms of the deviation of the proportion of that prey attacked from the proportion available in the environment. Such a simple definition belies the complex behavior upon which it may depend. For example, preference may result from differential searching rates, from different times spent in various habitat types, from active rejection of some prey types following their encounter, from differing abilities of prey to escape, and from any combination of these and other factors.

A variety of indices of preference have been amassed in the literature, many of which emerge as the same index in different guise. Their evolution, from the work of Scott (1920) to that of Jacobs (1974), is reviewed by Cock (1977, 1978). The analysis of preference considered here follows directly from the previous chapters since it depends upon the relative values for the functional response parameters, estimated separately for each prey type. This idea stems from Murdoch (1969, 1973), who suggested using an appropriate predation model, in this case a modified disc equation, to predict the extent of preference in the face of two prey types presented together. The number of each prey type eaten is predicted from

$$N_{a1} = \frac{a_1'TN_1}{1 + a_1'T_{h1}N_1 + a_2'T_{h2}N_2}$$
$$N_{a2} = \frac{a_2'TN_2}{1 + a_2'T_{h2}N_2 + a_1'T_{h1}N_1} \tag{7.1}$$

where the subscripts 1 and 2 distinguish the terms for the two prey species. The ratio of the two prey eaten is therefore given by

$$\frac{N_{a1}}{N_{a2}} = \frac{a_1'N_1}{a_2'N_2}, \tag{7.2}$$

where a_1'/a_2' is a measure of preference analogous to the index proposed by many other workers (cf. c in Murdoch, 1969).

Equation (7.1) suffers from the same disadvantage as the disc equation itself: it does not permit the exploitation of prey during the interval T. To overcome this, Lawton, Beddington, and Bonser (1974) and Cock (1977) make use of the random predator equation (3.6) to predict the numbers of each prey eaten:

$$N_{a1} = N_1[1 - \exp\{- a_1'(T - T_{h1}N_{a1} - T_{h2}N_{a2})\}]$$
$$N_{a2} = N_2[1 - \exp\{- a_2'(T - T_{h2}N_{a2} - T_{h1}N_{a1})\}].$$
(7.3)

The ratio of the two prey types eaten thus becomes:

$$\frac{N_{a1}}{N_{a2}} = \frac{N_1[1 - \exp(- a_1'T_s)]}{N_2[1 - \exp(- a_2'T_s)]},$$
(7.4)

where $T_s = (T - T_{h1}N_{a1} - T_{h2}N_{a2})$. (A similar procedure, based on the random parasitoid equation (3.5) would be adopted if parasitoids rather than predators were involved). The preference index is now considerably more complex than a_1'/a_2', being a variable dependent on the time spent searching T_s, which in turn depends on the numbers of both prey types eaten.

Using equation (7.3), Cock (1978) proposes the following convenient recipe for detecting preference.

(1) Carry out functional response experiments with each prey separately.
(2) Estimate a_1', a_2', T_{h1} and T_{h2} from the random predator (or random parasitoid) equation as described on page 34.
(3) Any preference resulting from differences in the functional response parameters (i.e., $a_1' \neq a_2'$ and/or $T_{h1} \neq T_{h2}$) can now be conveniently displayed in terms of N_{a1}/N_{a2} plotted against N_1/N_2 or, alternatively, as the

proportion of one of the species in the total diet against the proportion available (e.g. $N_{a1}/[N_{a1} + N_{a2}]$ against $N_1/[N_1 + N_2]$). Such *innate* preference will then be detected as a deviation from a slope of unity passing through the origin.

(4) Carry out a further experiment in which various ratios of the two prey types are presented together, and contrast predicted and observed results. Ideally, this procedure should then be repeated for a range of total prey densities that encompass those used in the functional response experiments (1). Any difference between the predicted preference from (3) and the observed preference will now be due either to an active rejection of one of the prey or to some change in a_1', a_2', T_{h1} or T_{h2} as a result of the predator experiencing the two prey types together.

An example of this procedure is given in Figure 7.1a and b using a mite predator-prey system. The separate functional responses to each prey are shown in Figure 7.1a and provide the estimates of the search rates and handling times needed to determine the innate preference shown by the line in Figure 7.1b. The fact that there is a deviation from the 45° line merely indicates that the two functional responses are not identical. The points in Figure 7.1b give the experimental results where both prey were presented together in varying ratios. They clearly conform well with the predicted results, indicating that the relative values of search rate and handling time from the separate functional responses fully account for the observed preference. But this by no means need be the case, as illustrated in Figure 7.2a and b. Here the observed results deviate markedly from the predictions, indicating that there is some change in predator searching behavior arising from both prey types being presented together (as in (4) above).

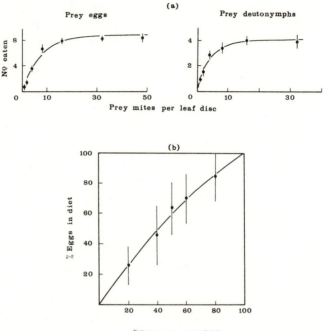

%Prey eggs available

FIGURE 7.1. (a) The functional responses of deutonymph *Phytoseiulus persimilis* where the prey are either eggs or deutonymphs of the red spider mite, *Tetranychus urticae*. The experiments were carried out over 24 hours with predators and prey confined to 16 cm² French bean leaf discs. The results are described by the "random predator" equation (3.6), with $a' = 2.73$ cm²/hr and $T_h = 2.55$ hrs for prey eggs, and $a' = 1.77$ cm²/hr and $T_h = 5.52$ hrs for prey deutonymphs. (Means and 95% confidence limits shown.) (b) The results of experiments in which eggs and deutonymph *T. urticae* were presented together in various ratios as prey for *P. persimilis*, using the same experimental conditions as in (a) above. The solid line shows the predicted ratios of prey eaten using equation (7.4) with search rates and handling times taken from the functional responses in (a). The points represent the mean ratios actually eaten (with 95% confidence limits), and show the observed preference to conform well with that predicted from the two functional response experiments (after Fernando, 1977).

126

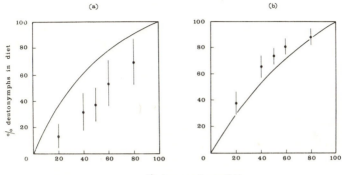

FIGURE 7.2. The results of further preference experiments of the kind displayed in Figure 7.1b, but now using different predator and prey stages. The solid lines are again the expected preferences obtained from the functional responses to each separate prey type. (a) Protonymph *Phytoseiulus persimilis* feeding on eggs and deutonymphs of *Tetranychus urticae* presented in the various ratios shown. The observed preference is for eggs over deutonymphs, and hence the reverse of that predicted. (b) Adult female *P. persimilis* feeding on larvae and deutonymphs of *T. urticae*. The observed preference is now more acute than that expected from the functional responses alone (after Fernando, 1977).

SWITCHING

The recipe outlined above also provides an ideal null hypothesis against which any tendency for "switching," *sensu* Murdoch (1969), may be tested. Switching implies that the proportion of a prey type taken changes from less than to greater than expected as the proportion of that prey available increases. A clear example of this is shown in Figure 7.3 from experiments by Lawton, Beddington, and Bonser (1974) using the predatory bug *Notonecta glauca* feeding on the mayfly, *Cloën dipterum*, and the isopod, *Asellus aquaticus*. Further examples from vertebrates and invertebrates are well reviewed by Murdoch and Oaten (1975), Murdoch, Avery, and Smyth (1975), Oaten and Murdoch (1975b), and Murdoch (1977).

127

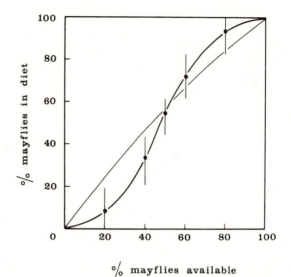

% mayflies available

FIGURE 7.3. The percentage of mayfly larvae *(Cloën dipterum)* in the diet of an adult *Notonecta glauca,* shown as a function of the availability of mayflies in relation to the alternative prey, *Asellus aquaticus.* Means and ranges from five replicates are shown, with the switching curve fitted by eye and the expected proportion of mayflies in the diet calculated using Murdoch's (1969) index: $c = 1.28$ (after Lawton, Beddington, and Bonser, 1974).

Switching is likely to be of most general importance in situations where the different prey types occur in somewhat disparate habitats, since it can now result simply from the allocation of a greater fraction of the searching time to whichever habitat is the more profitable in terms of biomass eaten (or hosts parasitized) per unit time. This is merely an extension of the aggregative responses discussed in Chapter 4, but now applied to different species.

Such a mechanism has been proposed by Royama (1970) as an alternative explanation to Tinbergen's (1960) development of a "specific searching image" to account for foraging by great tits. Royama assumes that predators allocate their searching time in relation to the profitability of a particular habitat in such a way that the total time per habitat is a rising function of prey density as shown by the two curves in Figure 7.4a. Whether these are convex *(A)* or sigmoid *(B)*, they both ultimately level off due to the combined effects of handling time and/or satiation. Just as a varying search efficiency a' was a cause of sigmoid functional responses in Chapter 3, so these relationships with varying time T also yield sigmoid responses when substituted into the random predator or parasitoid equations. This is illustrated in Figure 7.4b.

An experimental system that conforms well with this model is that of Murdoch, Avery, and Smyth (1975). These authors observed switching behavior in guppies feeding on a mixture of limbless, wingless *Drosophila* adults floating on the water surface and tubificid worms on the aquarium bottom. The guppies fed disproportionately on whichever prey was most abundant, spending increasing periods of time at the surface as the proportion of *Drosophila* increased.

Since polyphagous predators are likely to move readily

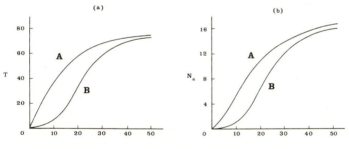

Prey density per habitat

FIGURE 7.4. (a) Two hypothetical curves relating the time spent in a particular habitat to the prey density in the habitat. (b) The corresponding sigmoid functional responses calculated from equation (3.5) with assumed values for a' and T_h, and the time available T being obtained from the corresponding curve in (a) (based on Royama, 1970).

from one prey type to another as prey abundances change, we may expect them to show particularly marked sigmoid functional responses to any given prey species. In Chapter 3 we saw that such sigmoid responses *alone* could not stabilize a discrete, coupled predator-prey model, due to the one-generation time delays between successive changes in the predator population density. When we turn away from such a coupled interaction to the generalist predators of this chapter, we have a system where the predator population fluctuations tend to be only loosely related, if at all, to the density of any one of their prey. By switching from prey species to prey species as their abundances change, the predator is effectively buffered against any vagaries in a part of its diet. The dynamics of such interactions between a generalist and just one of its prey will be completely different from those of the coupled system, largely because of the reduced time delays resulting from a more constant predator population. Thus the density dependence introduced by the sigmoid response can now act as a powerful stabilizing mechanism.

This can be conveniently illustrated using a model from Hassell and Comins (1978):

$$N_{t+1} = \lambda N_t \exp \left[-\frac{bN_t P_t}{1 + cN_t + bT_h N_t^2} \right]$$

$$P_{t+1} = P_t = P^*,$$

(7.5)

where the exponent stems directly from equation (3.10). This model has two potential equilibria, as shown in Figure 7.5, the lower of which (S) can easily be locally stable. If, however, the prey population exceeds the upper, unstable equilibrium (R), it "escapes" predator control and thenceforth increases exponentially. The local stability regions around the equilibrium (S) are illustrated in Figure 7.6 for two different values of the sigmoid functional response parameter c. Region B in the graphs represents locally unsta-

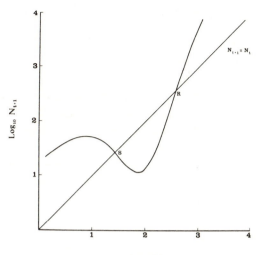

FIGURE 7.5. A population growth curve from equation (7.5) with $T_h = 0.01$, $b = c = 0.01$, $\lambda = 20$, and $P^* = 15$. The intersections with the 45° line, S and R, are respectively the lower potentially stable equilibrium and the upper unstable equilibrium (from Hassell and Comins, 1978).

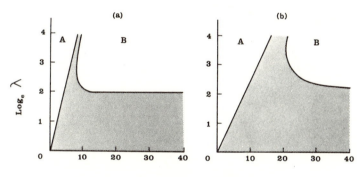

Predator density

FIGURE 7.6. Local stability boundaries in terms of the prey rate of increase λ and predator density P^* for the lower equilibrium S in Figure 7.5. The different boundaries in (a) and (b) show the effect of varying the sigmoid functional response parameters c, b and T_h. For ease of presentation, the predator density is scaled in units of $\sqrt{b/T_h}$, and similarly the parameter c is also scaled in terms of b and T_h: (a) $c = 0$; (b) $c = 2\sqrt{(bT_h)}$. The shaded areas show the domains of local stability and regions A and B are explained in the text (from Hassell and Comins, 1978).

131

ble equilibria, while in region A there are no equilibria at all since the curve in Figure 7.5 does not now intersect the 45° line. This model is qualitatively very similar to that of Southwood (1975, 1977c) and Southwood and Comins (1976), and Figure 7.5 is also reminiscent of similar figures in Takahashi (1964), Holling (1973), and Kiritani (1976).

The essential ingredient in model (7.5) is that the *total* response of the predators (measured in terms of N_a, the total prey eaten per generation) should show a sigmoid relationship with prey density. This, however, may be achieved in more than one way. In (7.5) it results from the combination of a sigmoid functional response and a constant predator density (i.e. no numerical response), as shown in Figure 7.7 (curves labeled B). An alternative, which is biologically perhaps more likely, is where the

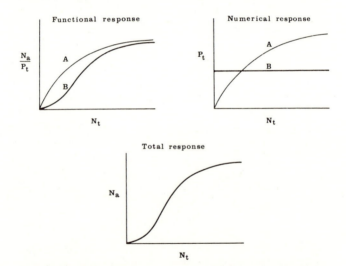

FIGURE 7.7. Schematic figures showing alternative ways of obtaining a sigmoid total response between the prey eaten by P_t predators, N_a, and prey density N_t. The total response curve may be obtained by combining either functional response A or B with numerical response A, or from functional response B without any numerical response.

number of predators searching for the prey is not constant, but tends to decline as that prey's density falls. A sigmoid total response is still obtained, however, when this is combined with either a type II or type III functional response (Figure 7.7, curves labeled *A*).

A DISCRETE TIME COMPETITION MODEL

As a first step in exploring the effects of predation in a multi-prey model we should have at hand a suitable interspecific competition model upon which to build. Such a model, a difference equation version of the Lotka-Volterra differential equations, has been described by May (1974) and Hassell and Comins (1976). It follows directly from the single species model (2.10b), and takes the form

$$X_{t+1} = \lambda X_t \exp[-g(X_t + \alpha Y_t)]$$
$$Y_{t+1} = \lambda' Y_t \exp[-g'(Y_t + \beta X_t)], \tag{7.6}$$

where g and g' are constants and α and β are the usual competition coefficients for the two species. This model retains the linear zero growth isoclines of the Lotka-Volterra model that separate the regions of positive and negative population growth ($X_{t+1} = X_t$; $Y_{t+1} = Y_t$), but differs fundamentally in the stability properties about the two-species equilibrium. Instead of always approaching the equilibrium monotonically, the time delays in the model also permit oscillatory damping, stable limit cycles, and apparent chaos (May, 1974, 1975b) as shown in Figure 7.8. A general stability diagram that shows the domains of these different behaviors is given in Figure 7.9. It is presented in terms of the potential equilibrium populations X^* and Y^*, since the same picture applies to *all* models of the form

$$X_{t+1} = X_t[f(X_t + \alpha Y_t)]^{-b}$$
$$Y_{t+1} = Y_t[f'(Y_t + \beta X_t)]^{-b'}, \tag{7.7}$$

of which model (7.6) is a special case where X^* and Y^* are given by the solution of the equations

$$\lambda = \exp[g(X^* + \alpha Y^*)]$$
$$\lambda' = \exp[g'(Y^* + \beta X^*)].$$
(7.8)

Thus the stability properties of the two-species equilibrium depend only on gX^*, $g'Y^*$, and the product $\alpha\beta$. The figure

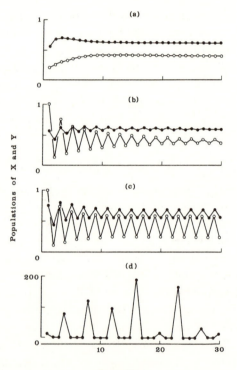

FIGURE 7.8. Numerical simulations from equation (7.6). The parameters used in (a), (b), and (c) correspond to points A, B, and C in Figure 7.9 and show the effect of gradually increasing the level of competition in species Y. (d) shows the apparently chaotic behavior occurring when competition is further increased. Note the compressed scale, and that species X has been omitted since its fluctuations are not visible on this scale (from Hassell and Comins, 1976).

134

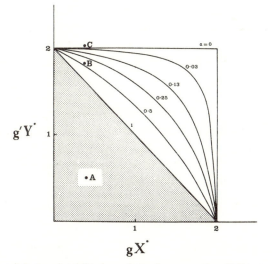

FIGURE 7.9. Local stability boundaries from equation (7.6) in terms of the potential equilibrium populations of the two species, X^* and Y^*. Equilibrium points falling within the relevant $\alpha\beta$ boundary (from 0 to 1) are stable; points outside the boundary correspond to limit cycles or chaotic behavior. The shaded area indicates the region of monotonic stability. No stable equilibrium is possible for $\alpha\beta > 1$. Points A, B, and C correspond to simulations (a), (b), and (c) in Figure 7.8 (from Comins and Hassell, 1976).

shows how the region of local stability varies with $\alpha\beta$ and also indicates (shaded area) the zone, independent of $\alpha\beta$, within which the convergence to the equilibrium is monotonic. (Note that there can be no stable two-species equilibrium if $\alpha\beta \geq 1$, corresponding to complete niche overlap.)

PREDATION IN MULTI-PREY COMMUNITIES

Our next step, to which this chapter has been leading, is to enlarge the competition model (7.6) by the addition of a predator population that attacks both prey species. In the first place this is done using the simplest kind of predator:

that envisaged by Nicholson and Bailey (1935). The model is then extended to include predators that show mutual interference, to predators that exhibit switching, and finally the whole model is broadened to an n-prey situation.

Random Predators

Let us consider a predator that searches randomly with a constant search rate (independent of prey and predator densities) for prey that it unfailingly consumes when encountered. This gives us the model

$$X_{t+1} = \lambda X_t \exp[-g (X_t + \alpha Y_t) - a_X P_t]$$
$$Y_{t+1} = \lambda' Y_t \exp[-g'(Y_t + \beta X_t) - a_Y P_t] \qquad (7.9)$$
$$P_{t+1} = X_t [1 - \exp(-a_X P_t)] + Y_t [1 - \exp(-a_Y P_t)],$$

where a_X and a_Y are the search rates per generation for the two prey species.

The inclusion of such a predator shifts the stability boundaries in a complex way. The major effects can be seen from the two examples in Figure 7.10 and from the simulations in Figure 7.11, and are summarized as follows.

(1) A three-species equilibrium is impossible if there is complete niche overlap between the competing prey ($\alpha\beta \geq 1$); one of the prey must inevitably move to extinction (Figure 7.11a), leaving a predator-prey interaction whose local stability properties are shown in Figure 2.7.

(2) As competition becomes less intense ($\alpha\beta \to 0$), the domains of stability for a three-species equilibrium are increased. Consider the case in Figure 7.10a where $\alpha\beta = 0.5$. Within the unhatched area, a stable equilibrium occurs, in part under conditions where the two prey species *alone* could only exhibit a locally unstable equilibrium. This is illustrated by the simulation in Figure 7.11b. Within the hatched area, however, the predator becomes extinct and

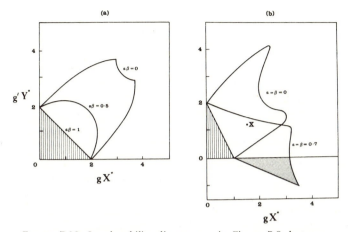

FIGURE 7.10. Local stability diagrams as in Figure 7.9, but now from equation (7.9) that includes predation. (a) With symmetric parameters: $a_X/g = a_Y/g' = 0.5$ and $\alpha/\beta = 1$. (b) With asymmetric parameters: $a_X/g = 1, a_Y/g' = 0.5$ and $\alpha/\beta = 1$. A stable equilibrium occurs within the unhatched areas. Within the hatched areas, the predator becomes extinct and the system collapses to a two-species competitive interaction described by equation (7.6). The shaded area denotes conditions where the two prey species alone could not coexist. The point X corresponds to the simulation shown in Figure 7.11b (after Comins and Hassell, 1976).

the system reverts to the two-species competition situation whose stability properties are given in Figure 7.9.

(3) Where preference for one of the prey species occurs (i.e. $a_X/a_Y \neq 1$), the interesting situation can arise in which predation is maintaining an equilibrium under conditions where the two prey species alone could not exist (Figure 7.10b, shaded area). The mechanism for this is illustrated in Figure 7.12. In the absence of predation, the zero growth isoclines (solid lines) do not intersect at all in the positive quadrant and hence the negative values for $g'Y^*$ in Figure 7.10b, indicating that one of the two prey species must move to extinction. But with the addition of a predator showing preference for the superior competitor, the

137

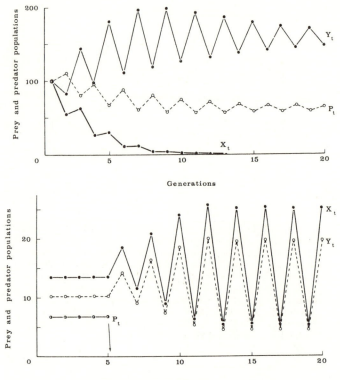

FIGURE 7.11. Numerical simulations from equation (7.9). (a) Prey X becomes extinct when $\alpha\beta > 1$; $a_X = a_Y = 0.008$, $g = g' = 0.01$, $\alpha = \beta = 1.1$, $\lambda = 2$, $\lambda' = 3$. (b) An example where predation is preventing limit cycle behavior on the part of the competing prey; $a_X = a_Y = 0.05$, $g = g' = 0.1$, $\alpha = \beta = 0.707$, $\lambda = 11$, $\lambda' = 10$.

two isoclines are shifted by different amounts (broken lines in Figure 7.12) which can now lead to a "cross-over" and hence a stable equilibrium point. Note that this effect is achieved by a displacement of the isoclines without any alteration in slope.

The significant conclusions to emerge from model (7.9) are therefore two-fold. (1) An equilibrium is impossible if

138

there is complete niche overlap ($\alpha\beta \geq 1$), and (2) randomly searching predators *can* maintain an otherwise unstable prey community, but only if preference for the different prey species delicately balances any competitive superiority. This does not contradict van Valen's (1974) view that "equivalent predation" cannot increase the numbers of prey species, since his "equivalent" predators search both randomly and without preference. These conclusions are not altered by introducing some stabilizing behavior into the basic predator-prey interaction, most easily done using the interference model (5.2), so that P_t in (7.9) is replaced by the term P_t^{1-m}. The major effect of this is to enlarge the regions of local stability in Figure 7.10, especially as competition between the prey species is made more intense ($\alpha\beta \to 1$). It does not alter the conclusion that there can be no equilibrium if $\alpha\beta \geq 1$.

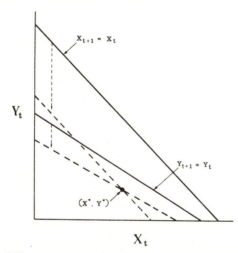

FIGURE 7.12. Zero growth isoclines for two competing prey species, X and Y. In the absence of predation, the isoclines do not intersect (solid lines). Predation with preference for one of the prey species depresses the isoclines by different amounts which can lead to a stable equilibrium point (broken lines) (from Comins and Hassell, 1976).

Switching Predators

An important qualitative difference emerges when the predators in model (7.9) are allowed to switch to whichever prey is the most abundant at that time. Now for the first time predators can maintain a stable equilibrium when $\alpha\beta \geq 1$. A convenient way to illustrate this is with the model

$$X_{t+1} = \lambda X_t \left[-g(X_t + \alpha Y_t) - (1 + E)a_X P_t \right]$$
$$Y_{t+1} = \lambda' Y_t \left[-g'(Y_t + \beta X_t) - (1 - E)a_Y P_t \right]$$
$$P_{t+1} = X_t \left[1 - \exp\{-(1 + E)a_X P_t\} \right]$$
$$\qquad\qquad + Y_t \left[1 - \exp\{-(1 - E)a_Y P_t\} \right], \tag{7.10}$$

where E has the following dependence on the relative prey abundance

$$E = s(X_t - Y_t)/(X_t + Y_t) \tag{7.11}$$

and s is a constant expressing the degree of switching (see Figure 7.13).

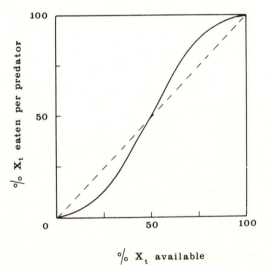

FIGURE 7.13. An example of predator switching from equations (7.10) and (7.11) where $a_X = a_Y = 0.5$ and $s = 1.0$.

140

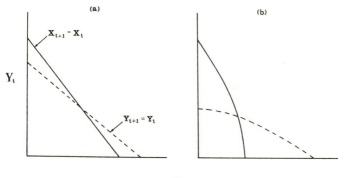

FIGURE 7.14. Zero growth isoclines for two prey species, X and Y: (a) without predation showing an unstable equilibrium: (b) with a switching predator, leading to curved isoclines and a stable equilibrium (from Comins and Hassell, 1976).

The crucial effect of the switching contained in model (7.10) is to make the prey zero isoclines become curved. This means that one can move from the necessarily unstable configuration shown in Figure 7.14a (obtained if $\alpha\beta \geq 1$), to the stable equilibrium in Figure 7.14b. Only a predator that switches can achieve this reversal of the zero isoclines.

Two examples of the stable parameter space that can result for different degrees of switching are shown in Figure 7.15. The expansion of the locally stable area as switching increases is seen to become increasingly apparent as $\alpha\beta$ increases. We may conclude, therefore, that a switching predator is very effective in stabilizing systems with strong interspecific competition, as indeed has been clearly shown by Roughgarden and Feldman (1975) for an extension of the Lotka-Volterra model with three competing prey and one predator species. On the other hand, such predators have only a modest effect on systems with weaker competition: for small $\alpha\beta$ there is only a small change in the area of stable space as switching increases.

141

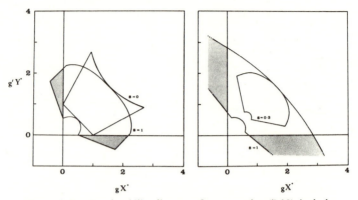

FIGURE 7.15. Local stability diagrams from equation (7.10), including predator switching *(s)* as in equation (7.11). (a) $a_X/g = a_Y/g' = 1$ and $\alpha = \beta = 0.45$; (b) $a_X/g = a_Y/g' = 1$ and $\alpha = \beta = 1.18$. The shaded area indicates where predation is stabilizing the system where otherwise not even a potential two-prey equilibrium could occur. Note that switching enlarges the stability boundaries, and that this effect is most striking when $\alpha\beta$ is large as in (b) (after Comins and Hassell, 1976).

Multiple Prey System

Certain of the effects of predators discussed in the previous sections require the system to be asymmetric; for example, asymmetry is required where preference permits an equilibrium point that could not exist in the corresponding two-prey system without predation (Figure 7.10b, shaded area). Nevertheless, the effects which expand or contract the zone of stability, rather than merely distort it, can be demonstrated in the entirely symmetric case (i.e. $\lambda = \lambda'$, $g = g'$, $\alpha = \beta$, $a_X = a_Y$), which can be simply extended to a system with n competing prey species:

$$(X_i)_{t+1} = \lambda X_i \exp\left[-g\left(X_i + \alpha \sum_{j\neq i} X_j \right) - (1 + E_i)aP_t \right] \quad (7.12a)$$

$$P_{t+1} = \sum_i X_i \left[1 - \exp\{-(1 + E_i)aP_t\} \right] \quad (7.12b)$$

142

The effects of switching are here defined by

$$E_i = s \ (X_i - \overline{X})/\overline{X}, \qquad (7.13)$$

where $\overline{X} = \left(\sum_i X_i\right)/n.$

Equation (7.12) represents an extreme case where all n prey species are competing "promiscuously." Alternatively, we could assume that the n prey species are spread out over a resource continuum (such as food size) with each species only competing with its "nearest neighbors" along the continuum. This assumption has frequently been made (see, for example, MacArthur, 1972; May and MacArthur, 1972; May, 1975a, 1976c) and requires that the sum over j in equation (7.12a) be restricted to $j = i - 1, j = i + 1$. (Note that in order to preserve symmetry in this case, it is necessary to increase g for each of the end species, $i = 1$ and $i = n$).

The detailed stability analysis of these models is given by Comins and Hassell (1976), who found that the conclusions from the two-prey models remain generally intact. Thus it remains true that stability is impossible if there is complete niche overlap, as long as there is no predator switching. This is illustrated in Figure 7.16a where the stable zone is seen to diminish as α increases, finally disappearing when $\alpha = 1$. Only by introducing a switching predator, as in Figure 7.16b, does it become possible to retain some stable space with $\alpha \geq 1$.

It is now fitting to revisit briefly Paine's (1966) experiment on the exclusion of *Pisaster*, in which several of the prey became extinct and two of them, *Mytilus* and *Mitella*, came to dominate the system. Subsequent work by Landenburger (1968) and Paine (1974) gives some valuable insights into the system. Landenburger has shown that *Pisaster* in the laboratory has a clear *preference* for *Mytilus*, and Paine provides support for *Pisaster* being able to *switch*

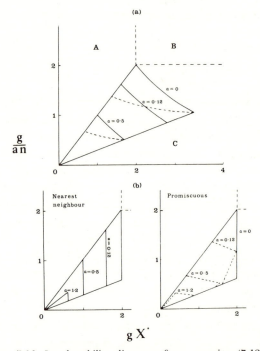

FIGURE 7.16. Local stability diagrams from equation (7.12); (a) without predator switching ($s = 0$), (b) and (c) with predator switching ($s = 1$). The solid lines show the boundaries for "nearest neighbor" competition and the broken lines for "promiscuous" competition for various values of α. In region A the predator is eliminated leaving a stable n-prey species equilibrium. The predator is also eliminated in region B, but now there remains a locally unstable n-prey equilibrium. In region C there is a locally unstable predator-prey equilibrium and higher order cyclical behavior occurs (after Comins and Hassell, 1976).

to other prey types when *Mytilus* are scarce. Unfortunately, similar information for *Mitella* is not available. As we have seen, these are both important behaviors in enabling a predator to maintain an otherwise unstable community of prey. Preference can enhance coexistence when $\alpha\beta < 1$, while sufficient switching can maintain a prey community with complete niche overlap.

144

As a final note to this chapter, we return to switching predators and a résumé of what they can and cannot do in predator-prey systems. In a trenchant note on this, May (1977) regrets that "too often, simple accounts portray switching as a kind of Universal Stabilizer." That this is not necessarily the case was first emphasized by Steele (1974). Using a differential equation system, he showed that switching can only stabilize a system in which the individual predator-prey links are themselves potentially stable. The same point has recently been made by Oaten and Murdoch (1977) and also holds true for the discrete generation systems discussed in this chapter. Thus, without the density dependent prey growth rates in equations (7.10) or (7.12) there could be no stable three-species or $(n + 1)$-species equilibrium respectively, however much switching were included.

The models in this chapter emphasize a further point: switching has its most pervasive effect when competition between the prey species is most severe (i.e. when $\alpha\beta \geq 1$). This is clear from the stability diagrams in Figure 7.15; switching has little effect in enhancing conditions for stability when $\alpha\beta \ll 1$, but a considerable effect as soon as complete niche overlap is approached. Finally, we should note that a *fixed preference* by the predator for a superior competitor can, like switching, maintain an equilibrium where the prey alone have no equilibrium. This differs from the effects of switching, however, in not applying whenever $\alpha\beta \geq 1$.

SUMMARY

Polyphagous predators are likely to exhibit some degree of preference for one or more of their prey species. Means of quantifying such preference are discussed and a recipe for detecting preference from laboratory experiments is outlined.

Switching differs from a fixed preference for a particular prey species in that the proportion of a prey type attacked changes from less than to greater than expected as the proportion of that prey available increases. Such switching is likely to lead to sigmoid functional responses for a given prey species, which can contribute markedly to the stability of that prey population as long as the predator population remains relatively constant.

Both preference and switching by "top predators" can have a significant effect on the coexistence of competing prey species. This is illustrated by a discrete generation predatory-prey model with two or more prey. The predators are allowed either to search randomly, with or without preference for one of the prey, or allowed to switch to whichever prey is the most abundant at the time. Both preference and switching are important factors in enabling a polyphagous predator to maintain an otherwise unstable prey community.

CHAPTER EIGHT

Competing Predators and Hyperparasitoids

Almost all phytophagous arthropods are attacked by more than one, and often by many, predator species. Typically, the majority of these will be polyphagous, although among the parasitoids in particular there are likely to be some that are relatively specific to a particular host species. It is convenient, therefore, to refer primarily to parasitoids in this chapter, because the population models are yet again to be framed as coupled difference equations, and because parasitoids provide the best known examples from the real world. The chapter falls into two distinct parts. First, we pose the question, what conditions permit a single host species to support more than one parasitoid species in a stable interaction? We then turn to a rather different system, where the primary parasitoid species is itself attacked by another parasitoid, namely, a secondary or hyperparasitoid. Once again, we seek some understanding of the conditions permitting the system to persist.

SEVERAL PARASITOIDS-ONE HOST

An excellent review of the structure of parasitoid complexes attacking phytophagous hosts is given by Zwölfer (1971). He notes that field populations, of moths and sawflies in particular, often support numerous parasitoid species, ranging from those which tend to be fairly specific to a single host species, to others than are widely polyphagous (see Table 8.1). Characteristically, the specialists are in

147

a minority, but this classification of host range may often be misleading. Zwölfer lists three categories of parasitoids that may be *effectively* quite specific to a given host.

(1) There are, of course, some complete specialists. These are relatively rare and indeed absent from many natural parasitoid complexes.

(2) There are oligophagous species which tend to concentrate on a single host species.

(3) There are some notorious generalists that may in some locations exhibit fairly specific behavior. A good example comes from the ichneumonid *Itoplectis conquisitor* which in some regions has become adapted to two hosts that have recently been introduced to Canada. Perhaps the associative learning by female *I. conquisitor* demonstrated by Arthur (1966) provides the underlying explanation.

To model such complex systems in any detail would force us to abandon any thoughts of analytical models in favor of simulation. Rather than do this, we shall follow the pattern of previous chapters and explore the simplest case—now with one host *(N)* and two parasitoid species *(P* and *Q)*—in the hope that useful general conclusions will emerge.

Let us consider the general model

$$N_{t+1} = \lambda \, N_t f_1(P_t) f_2(Q_t)$$
$$P_{t+1} = N_t \, [1 - f_1(P_t)] \qquad (8.1)$$
$$Q_{t+1} = N_t f_1(P_t) \, [1 - f_2(Q_t)],$$

where the function f_1 is the probability of a host not being found by P_t parasitoids and similarly f_2 is the probability of not being found by Q_t parasitoids. This kind of model applies to two types of host-parasitoid interaction, both of which yield exactly the same end result and are frequently to be found in real systems. It applies to cases where P acts first, to be followed by Q that acts on the surviving hosts. Such is the situation where the host species is attacked in different developmental stages by a range of parasitoids. The winter moth in Table 8.1, for example, is parasitized by egg, larval, and pupal parasitoids. It also applies to cases where P and Q act simultaneously on the same host stage, with the larvae of P always out-competing those of Q should multiparasitism occur. This is depicted schematically in Figure 8.1. For instance, of the two parasitoids of winter moth larvae introduced into Canada (see Table 8.1), *Agrypon* larvae always eliminate *Cyzenis* when within a single host. *Cyzenis* can therefore be viewed as effectively searching only for the hosts surviving from *Agrypon*. Thus, only where the outcome of multiparasitism depends in some measure on the order of arrival within the host (e.g. Anderson, Hoy, and Weseloh, 1977) is a different model structure from that in (8.1) required.

The first analysis of equation (8.1) is due to Nicholson (1933) and Nicholson and Bailey (1935), who assumed that

$$f_1(P_t) = \exp(-a_1 P_t)$$

and

$$f_2(Q_t) = \exp(-a_2 Q_t),$$

(8.2)

TABLE 8.1. Range of specificity of the parasitoids of the winter moth (*Operophtera brumata*) in Europe and Canada. (After Zwölfer, 1971; details in Sechser, 1970).

	Europe	*Canada*
Host native or introduced?	Native	Introduced
Status of parasitoid species:		
Widely polyphagous	16 spp. (N)	19 spp. (N); 1 sp. (I)*
Moderately polyphagous	8 spp. (N)	none
Fairly specific	2 spp. (N)	1 sp. (I)**

N.B. (N) and (I) denote native and introduced parasitoids respectively. The introduced species are *Agrypon flaveolatum* (*) and *Cyzenis albicans* (**).

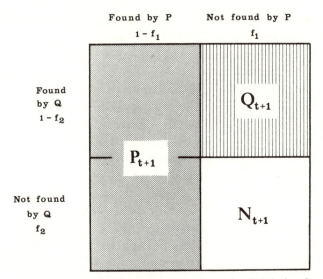

FIGURE 8.1. One of the interpretations of equation (8.1) is that the host species, N, is attacked by both parasitoid species, P and Q, at the same time, but that P is the superior larval competitor in cases of multiparasitism. This is shown schematically in the diagram. All hosts found by P (shaded area) give rise to P in the next generation (P_{t+1}), whether or not they have also been found by Q. Q_{t+1} (hatched area) arises only from hosts found by Q alone.

where a_1 and a_2 are the search rates per generation (i.e. areas of discovery) for P and Q. They found that for given values of a_1 and λ, the host rate of increase, there is a narrow "window" of a_2-values that permit an equilibrium, albeit one that is always unstable. Working with such an unstable model enables few conclusions to be drawn about the real world. The really interesting questions concern the range of conditions that enable a *stable* equilibrium to be achieved. Is it possible, for instance, to have species P primarily responsible for a depression of the prey equilibrium and Q contributing mainly to stability? Such questions are of direct relevance to biological control (see Chapter 9) as well as to an understanding of natural communities.

That a stable multi-species assembly is possible with equation (8.1) was shown by Hassell and Varley (1969). They used their modification of the Nicholson-Bailey model (equation (5.2)) to simulate a system with three parasitoids acting in sequence, as shown in Figure 8.2. In this chapter, we shall look for more general conclusions using a two parasitoid-one host model based on equation (4.19). Thus f_1 and f_2 of equation (8.1) become

$$f_1(P_t) = \left[1 + \frac{a_1 P_t}{k_1}\right]^{-k_1}$$

$$f_2(Q_2) = \left[1 + \frac{a_2 Q_t}{k_2}\right]^{-k_2},$$

(8.3)

where k_1 and k_2 are the exponents of the negative binomial distribution describing the effects of parasitoids aggregating in patches of high host density, as discussed on page 74.

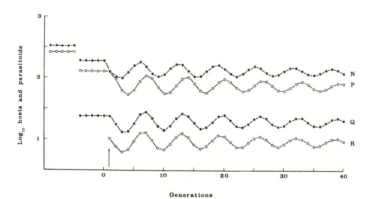

FIGURE 8.2. A numerical simulation from a model with a host (N) and three parasitoid species $(P, Q,$ and $R)$ acting in sequence. The model is an extension of equation (8.1) to include a third parasitoid species (R), and the functions for parasitism—f_1, f_2, and f_3—are defined by the interference sub-model, $f = a P_t^{1-m}$, as in equation (5.3). The parameter values used are $\lambda = 5$, $a_p = a_q = a_r = 0.5$, $m_p = m_q = m_r = 0.5$. To the left of the figure are shown the equilibrium populations of N and P alone, and for N, P, and Q. Parasitoid R is then introduced at the point marked by the arrow.

Consequently, with $k_1 = k_2 = \infty$, the three-species Nicholson-Bailey model is reobtained.

The analysis of model (8.1) with f_1 and f_2 from equation (8.3) is given in May and Hassell (1979). They show that, given

$$g_1(k_1, \lambda) = \frac{\lambda - 1}{k_1(\lambda^{1/k_1} - 1)} \tag{8.4}$$

and

$$g_2(k_2, \lambda) = \frac{\lambda\, k_2(\lambda^{1/k_2} - 1)}{\lambda - 1}, \tag{8.5}$$

then the existence of a three-species equilibrium requires that either

$$g_1 > \frac{a_2}{a_1} > g_2 \tag{8.6}$$

or

$$g_1 < \frac{a_2}{a_1} < g_2. \tag{8.7}$$

Criterion (8.6) always results in an unstable equilibrium and is therefore of only passing interest here. Much more important is that of (8.7), since therein lie conditions where a three-species stable equilibrium may result. This is illustrated in Figures 8.3 and 8.4 for three special cases to be considered in turn: (1) where $k_1 = k_2 = k$, (2) where $k_1 = k$ and $k_2 \to \infty$, and (3) where $k_1 \to \infty$ and $k_2 = k$.

(1) $k_1 = k_2 = k$

The equilibrium conditions for this example lie between the upper and lower curves in Figure 8.3 (assuming $\lambda = 2$, as we do throughout). The arrows to the right of the figure show the asymptotic values of g_1 and g_2 where $k = \infty$, and therefore illustrate the narrow "window" of equilibrium conditions referred to above for the Nicholson-Bailey

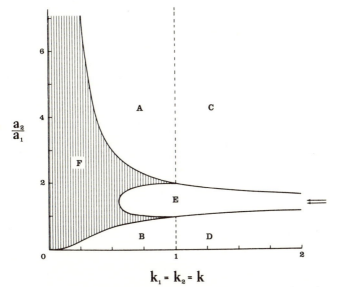

$$k_1 = k_2 = k$$

FIGURE 8.3. Local stability boundaries for equation (8.1) where f_1 and f_2 are defined in equation (8.3) and k_1 and k_2, the exponents of the negative binomial distribution, are equal. The upper and lower boundaries, g_2 and g_1, are defined from equations (8.4) and (8.5). Between these limits a three-species equilibrium is possible, but only within the hatched area (F) is it locally stable. The stability properties within the six regions, A to F, are described in the text. The arrows to the right of the figure indicate the asymptotic values of the boundaries (i.e. $g_1 = (\lambda - 1)/\log_e\lambda$ and $g_2 = (\lambda \log_e\lambda)/(\lambda - 1)$ and correspond to the "window" of equilibrium conditions for the three-species Nicholson-Bailey model referred to in the text.

model. For smaller values of k, and in particular for $k < 1$, this equilibrium space enlarges considerably, with the hatched area denoting the conditions for a locally stable three-species system to exist. The rather intricate stability diagram that results is best understood by demarcating six areas in Figure 8.3.

(A) and (B). Either species P (area A) or Q (area B) becomes extinct, leaving the other in a two-species interaction that is inevitably stable, since $k < 1$.

153

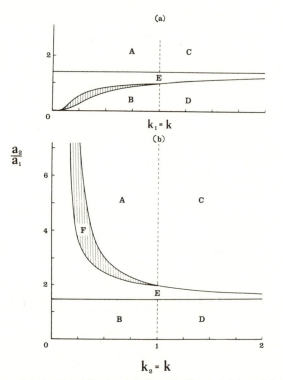

FIGURE 8.4. Local stability boundaries as in Figure 8.3, but with (a) $k_1 = k$, $k_2 \to \infty$ and (b) $k_1 \to \infty$, $k_2 = k$.

(C) and *(D)*. As above, but the interaction is now always unstable, since $k > 1$.

(E). A three-species equilibrium exists, but one that is always locally unstable.

(F). In this hatched area both P and Q can persist in a stable, three-species interaction. This equilibrium, however, is only locally stable with the domains of attraction being very large when $k \ll 1$ and progressively decreasing as $k \to 1$. The two sharp "horns" of the stable area arise where the stability is largely due to just one of the

154

parasitoids, with the other being but a small addition with a very low equilibrium population. The interaction is therefore approaching that of the single parasitoid-host system in the adjacent area, A or B.

$$(2)\ k_1 = k;\ k_2 \to \infty$$

This is the situation where species P can respond to host density per patch but Q always searches randomly. The equilibrium conditions are now much restricted, as seen from Figure 8.4a, and include only a small area of locally stable space. Regions A to F have the same interpretation as in Figure 8.3 except that the two-species interaction in area A is unstable.

$$(3)\ k_1 \to \infty;\ k_2 = k$$

Species P now searches randomly, with Q showing aggregated search. The equilibrium and stability conditions are shown in Figure 8.4b, with area B now representing an unstable two-species interaction.

We may conclude from these examples that a multi-parasitoid system is most likely to exist where all the parasitoid-host links are potentially stabilizing (i.e. $k < 1$ for all species). The possibilities for coexistence become very much more restricted whenever one or more of the links are destabilizing, as evident here when P or Q are allowed to search randomly.

It is also apparent from Figure 8.3 that provided $k \ll 1$, the locally stable area is more extensive when species Q has the larger of the two search rates (i.e. where $a_2/a_1 > 1$). Stated more fully: there are more possibilities of a stable equilibrium if the *inferior* larval competitor (i.e. species Q) has the larger search rate. That this is frequently so has been emphasized by Zwölfer (1971) who gives an impressive array of examples. Some of these are shown here in

TABLE 8.2. Some examples of "counter-balanced" competition between parasitoid species attacking the same host. (After Zwölfer, 1971).

Level of Competition	Mechanism	Superior species	Inferior species	Host species
Larval competition within host (i.e. "intrinsic competition")	Territorial behavior of 1st instar larvae eliminates competitor	*Temelucha interruptor*	*Orgilus obscurator*	Pine shoot moth (*Rhyacionia buoliana*)
	Competitor eliminated by direct feeding	*Itoplectis maculator*	*Cephaloglypta murinanae*	European fir budworm (*Choristoneura murinana*)
	Competitor eliminated by faster development and consumption of host	*Aptesis abdominator*	*Cyzenis albicans*	Winter moth (*Operophtera brumata*)
Adult efficiency in exploiting host population (i.e. "extrinsic competition")	Differences in searching efficiency	*O. obscurator*	*T. interruptor*	As above
	Difference in synchrony with host	*C. murinanae*	*I. maculator*	
	Differences in egg complement	*C. albicans*	*A. abdominator*	

TABLE 8.3. The relationship between larval competitive ability and the adult reproductive capacity in three species of parasitoids attacking *Spodoptera praefica*. The average eggs per adult female of each species are shown in parentheses. (After Miller, 1977).

Inferior larval competitor		Superior larval competitor	
Chelonus insularis	(450)	*A. marginiventris*	(210)
C. insularis	(450)	*Hyposoter exiguae*	(130)
Apanteles marginiventris	(210)	*H. exiguae*	(130)

Table 8.2. Further evidence comes from Miller's (1977) study on three species of parasitoids attacking *Spodoptera praefica*. Table 8.3 shows that for all three combinations of multiparasitism, the inferior larval competitor has the higher potential fecundity as an adult which, although not strictly a measure of searching efficiency, may well scale in the same way.

The effect of a second parasitoid species is more far-reaching than is revealed by a stability analysis alone; it can also markedly depress the equilibrium level of the host population. This is of such central importance to the practice of biological control that its discussion is better deferred to the next chapter, and we instead turn now to a rather different kind of three-species interaction.

HOST-PARASITOID-HYPERPARASITOID SYSTEMS

An interesting feature of insect parasitism is the frequent occurrence of secondary or hyperparasitism, where a parasitoid seeks out the immature stage of another parasitoid (the primary) as its host. Detailed information on the behavior and ecology of these hyperparasitoids is relatively sparse compared to the voluminous literature on primary parasitoids. Some notable contributions are the studies on aphid and scale insect hyperparasitoids by Weseloh (1969), Gutierrez (1970a-d), Gutierrez and van den Bosch (1970a, b), and Sullivan and van den Bosch (1971). Why hyperparasitoids have generally been so neglected is unclear. As Zwölfer (1971) states, "The role played by hyperparasites varies greatly in the different parasite complexes. Some primary parasites remain remarkably free of hyperparasites whilst others suffer heavy losses. So far, hyperparasites have been largely neglected in studies on the dynamics of host-parasite systems, but available information suggests that their influence on the reproductive efficiency of primary parasites is a major factor in some parasite complexes."

While most hyperparasitoid life cycles follow a straightforward pattern, some exhibit amazing variation. Flanders (1959), for instance, reports on species of *Coccophagus* where the females are always primary parasitoids of scale insects, while the males develop hyperparasitically on another parasitoid of the scales. Some other hyperparasitoids show a seemingly more adaptive behavior: they are *facultative* hyperparasitoids which can develop either as primary or secondary parasitoids, depending on whether the host found is healthy or already parasitized. Askew (1961), for instance, describes the complex food webs that occur within oak galls formed by cynipid wasps, where it is common to find several facultative hyperparasitoids among the parasitoid community. Muesebeck (1931) gives a further example: *Monodontomerus aereus,* a chalcidoid wasp that is either a primary parasitoid of brown-tail or gypsy moth pupae, or a hyperparasitoid attacking one of the several primary parasitoids of these moths.

These facultative hyperparasitoids provide an excellent intermediate situation between the two-parasitoid system of the previous section and the normal host-parasitoid-hyperparasitoid interaction. The relevant population

158

model is given by

$$N_{t+1} = \lambda\, N_t f_1(P_t)\, f_2(Q_t)$$
$$P_{t+1} = N_t f_2(Q_t)\, [1 - f_1(P_t)] \qquad (8.8)$$
$$Q_{t+1} = N_t\, [1 - f_2(Q_t)],$$

where Q is now the facultative hyperparasitoid and f_1 and f_2 are exactly as in equation (8.3). This, therefore, represents an interaction where species Q acts as a hyperparasitoid on encountering a host with species P within, but acts as a primary parasitoid whenever a healthy host is found (see Figure 8.5). A hyperparasitoid of this kind thus turns out to obey a model *identical* with a two-parasitoid system in which species Q wins over species P in all cases of multiparasitism. This means that equations (8.1) and Figure (8.1) apply once more, provided that we interchange P and Q, and f_1 and f_2. It is merely a matter of christening which is the "first" and which the "second" parasitoid species. Simi-

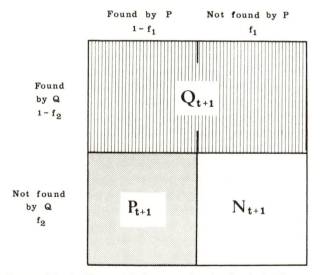

FIGURE 8.5. As Figure 8.1, but now for the facultative hyperparasitoid system of equation (8.8).

159

larly, the appropriate stability boundaries are the same as in Figures 8.3 and 8.4 provided that a_1 and a_2, k_1 and k_2, and P and Q are all interchanged.

The more usual host-parasitoid-hyperparasitoid interaction follows as the next step, in which the hyperparasitoids (species Q) can no longer reproduce in *any* host individual found, but only in those parasitized by species P (see Figure 8.6). Attempts at modeling such a system have been few. Nicholson and Bailey (1935) simulated a host-parasitoid-hyperparasitoid interaction which was unstable, but only Beddington and Hammond (1977) have given an analytical treatment. They adopt a model with a density dependent host rate of increase of the form of equation (2.10a) and with both primary and hyperparasitoid searching in a Nicholson-Bailey manner. Their model is therefore an extension of Beddington, Free, and Lawton's (1975) model (2.12) of a single parasitoid-host interaction. Bed-

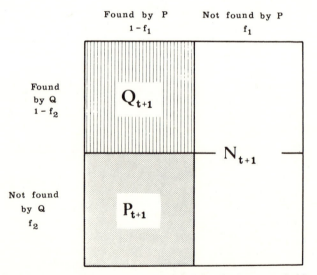

FIGURE 8.6. As Figures 8.1 and 8.5, but for the host-parasitoid-hyperparasitoid interaction of equation (8.9).

dington and Hammond's treatment is comprehensive, giving the conditions for a locally stable three-species equilibrium and a clear exposition of how the introduction of a hyperparasitoid affects the host equilibrium.

In this section, however, rather than rely on the host's density dependence for stability, we follow in the mold of equation (8.1) and (8.8) to give the model

$$N_{t+1} = \lambda N_t f_1(P_t)$$
$$P_{t+1} = N_t [1 - f_1(P_t)] f_2(Q_t) \qquad (8.9)$$
$$Q_{t+1} = N_t [1 - f_1(P_t)] [1 - f_2(Q_t)],$$

where f_1 and f_2 are again given in equation (8.3). Once again the detailed analysis is outlined in May and Hassell (1979) and only the outcome discussed here.

The condition for a three-species equilibrium to exist are two-fold. First, when $k_2 > 1$, an equilibrium exists only if

$$\frac{a_2}{a_1} < \frac{1}{k_1 (\lambda^{1/k_1} - 1)}, \qquad (8.10)$$

but proves *never* to be stable. If, however, $k_2 < 1$, then an equilibrium exists when

$$\frac{a_2}{a_1} > \frac{1}{k_1 (\lambda^{1/k_1} - 1)}, \qquad (8.11)$$

in which case it may or may not be stable depending on the value of λ and k_1.

A convenient situation to be explored is where $k_1 = k_2$, which leads to the local stability boundaries shown in Figure 8.7. The condition (8.11) is satisfied above the line and a locally stable three-species equilibrium exists throughout the hatched area, ending abruptly at $k_1 = k_2 = 1$.

The conclusion that emerges once more is that the system will be locally stable as long as the individual links are themselves stabilizing—that is, provided that k_1 and k_2 are less than unity. Figure 8.7 also emphasizes, much as to be

TABLE 8.4. Examples where search rates have been calculated for both primary and hyperparasitoids.

Host Species	Primary Parasitoid	Hyperparasitoid	a_1	a_2	a_2/a_1	Authors
Operophtera brumata (winter moth)	Cyzenis albicans	Phygadeuon dumetorum	0.075 m²/gen	1.68 m²/gen	22.4	Hassell (1969); Kowalski (1977)
Coccus hesperidum (soft brown scale)	Microterys flavus	Marietta exitiosa	0.081*	0.272*	3.4	Kfir, Podolev and Rosen (1976)
Coccus hesperidum (soft brown scale)	Microterys flavus	Cheiloneurus paralia	0.081*	0.354*	4.4	Kfir, Podolev and Rosen (1976)
Pristiphora erichsonii (larch sawfly)	Olesicampe benefactor	Mesochorus dimidiatus	0.35 m²/gen	0.35 m²/gen	1.0	Ives (1976)

* A laboratory study in which a_1 and a_2 have units of cage size per 48 hours.

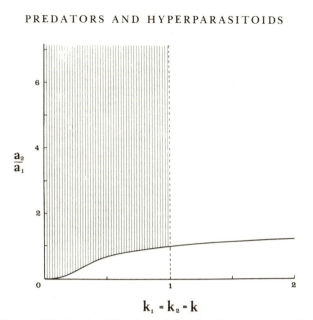

k₁ = k₂ = k

FIGURE 8.7. Local stability boundaries for the host-parasitoid-hyperparasitoid system of equation (8.9) where $k_1 = k_2$. The lower boundary, above which a three-species equilibrium is possible, is defined in equation (8.11). The hatched area shows where a locally stable three-species equilibrium occurs.

expected, that stability occurs mainly in the region where the hyperparasitoid has a larger searching efficiency than the primary (i.e. where $a_2/a_1 > 1$). The available information on this is summarized in Table 8.4 and supports the conclusion rather well.

As with the two-parasitoid system of the previous section, the effect of hyperparasitism on the equilibrium level of the host is important to biological control and hence best deferred to the next chapter.

SUMMARY

This chapter has dealt with two types of parasitoid-host system: one with two parasitoids and one host species, and

163

the other a host-parasitoid-hyperparasitoid interaction. Both have been modeled using coupled difference equations where the function for parasitism has taken the form of the detail independent model of predator aggregation from Chapter 4.

The two-parasitoid model applies to situations where either the parasitoids act successively or, equivalently, where they act simultaneously but one species always wins in larval competition within a host. The model is most likely to have a locally stable three-species equilibrium if one or both of the parasitoid-host links are potentially stabilizing (i.e. k_1 and $k_2 < 1$) and when the "second" parasitoid species has the higher of the two search rates (i.e. $a_2/a_1 > 1$). It is encouraging, therefore, to find several field examples where the inferior larval competitor (the "second" parasitoid) apparently has the higher searching efficiency.

A similar analysis has been applied to a host-parasitoid-hyperparasitoid system and shows again that the interaction is likely to be locally stable as long as both links are potentially stabilizing and if the "second" species, the hyperparasitoid, has the higher searching efficiency. This too is supported by some real examples.

CHAPTER NINE

A Theoretical Basis for Biological Control

Biological control has emerged as a powerful pest control technique during the last hundred years or so. Its origins, however, go back much further, at least to about 400 AD when the Chinese are known to have placed nests of the ant *Oecophylla smaragdina* in their citrus trees to control the citrus stink bug and other pests, a practice that reportedly continues today (Pu, 1976).

In classical biological control programs against exotic pests, natural enemies are generally sought in the general region where the pest originated, and are then imported and screened under quarantine before being bred in large numbers prior to release. It is hoped that the natural enemy population will increase rapidly after release and cause the pest population to decline, to be followed by both natural enemy and pest populations persisting at very low densities. Examples of such successes are plentiful and reviewed in several recent texts on biological control (Huffaker, 1971; van den Bosch and Messenger, 1973; DeBach, 1974; Huffaker and Messenger, 1976). It is clear from these that pests of such standing crops as fruit and forest trees are particularly amenable to biological control, while programs against pests of annual crops have been less successful (see Table 1 of Southwood, 1977c). The reason for different levels of success in the different kinds of habitat is not hard to find: the perennial standing crop permits a continuous interaction between natural enemy and host without the ecological upheavals associated with the regu-

165

lar harvesting and plowing in an annual field crop system. It is thus to the perennial standing crop systems that we should turn in attempting to relate the ideas of the previous chapters to biological control practices.

The majority of spectacular successes in the biological control of insect pests have come from parasitoids rather than predators. The most notable exception that springs to mind is that of the cottony cushion scale, *Icerya purchasi*, controlled by the coccinellid beetle, *Rhodolia cardinalis* (Koebele, 1890). *Rhodolia*, however, is an unusual predator in that certain features of its life cycle make it "parasitoid-like"; in particular, the adult females lay their eggs beside or under scales, each one of which will provide sufficient food for complete larval development.

A common explanation for the generally poor showing of predators in biological control programs is their supposed tendency to be more polyphagous than parasitoids. This would prevent the closely coupled interaction required to maintain low equilibrium populations. An additional feature of predators that may be important was discussed in Chapter 6, namely, their threshold prey consumption below which they cannot reproduce. Because of this, a coupled predator-prey model will have different local and global stability properties, with the result that an equilibrium might not be achieved simply due to an inappropriate initial ratio of predators and prey (Beddington, Free, and Lawton, 1976). A possible example of this phenomenon is the biological control of red spider mite in glasshouses by the predatory mite, *Phytoseiulus persimilis*, in which the initial predator-prey ratio is known to be crucial (Hussey and Bravenboer, 1971).

While the introduction of natural enemies in a biological control project will always be tinged with uncertainty, ecological theory can now contribute much to the development of a general theory for biological control practices. It identifies the factors necessary for a reduced and stable pest equilibrium, provides insights into the use of specific and polyphagous natural enemies, and suggests that multiple introductions of natural enemy species is a sensible strategy. This chapter is devoted to these ideas.

EQUILIBRIUM LEVELS

Most definitions of biological control make some explicit statement about a desired reduction in pest equilibrium levels. This is the primary objective, to which we should add the need for the interaction to remain sufficiently stable to prevent the host from sporadically re-emerging as a pest. Some features of an interaction leading to reduced equilibria are discussed in this section and those contributing to stability in the next.

First, let us consider again one of the simplest of parasitoid-host models, that of Nicholson and Bailey (2.5). The equilibrium here depends solely on λ and a, as shown in Figure 2.2. The overriding problems in relating this picture to the practice of biological control lie in the oversimplified behavior of the predators and, at least as important, in the total neglect of the many hazards that affect both host and parasitoid populations. It is quite unrealistic, for example, to assume that a host population suffers mortalities only from parasitism, and that the parasitoids are themselves free from mortality. Thus, the λ-axis of Figure 2.2 should become the effective host rate of increase after all factors other than parasitism have been considered, and the a-axis should become the overall performance of the parasitoid after due allowance has been made for all factors affecting parasitoid search and the survival of parasitoid progeny (Hassell and Moran, 1976; Hassell, 1977). A convenient measure of this "overall perfor-

167

mance," A, at least in homogeneous environment, is given by

$$A = \frac{1}{P_t} \log_e \left(\frac{N_t}{N_t - P_{t+1}} \right). \tag{9.1}$$

Comparison of this with equation (2.7) shows that A and a from the Nicholson-Bailey model will be identical when there is complete survival of parasitoid progeny, but will deviate as soon as the number of hosts parasitized differs from the numbers of parasitoid adults in the next generation (i.e. when $P_{t+1} \neq N_a$). (This is equivalent to c in equation (1.1b) being less than unity.)

A most graphic example of how these two measures, A and a, can differ comes from the well-known studies of the winter moth, *Operophtera brumata*, in Wytham Wood, England (Varley and Gradwell, 1968; Varley, Gradwell, and Hassell, 1973) and in Nova Scotia, Canada (Embree, 1965, 1966, 1971). What makes this such an interesting case study is the notable contrast between the population dynamics in the two locations. The winter moth in Nova Scotia, after its initial accidental introduction, spread rapidly and became a serious defoliator of hardwood trees during the early 1950s. A biological control program was initiated in Nova Scotia in 1954 in which two of the released parasitoids, a tachinid *Cyzenis albicans* and an ichneumonid *Agrypon flaveolatum*, became established. Parasitism increased rapidly, with *Cyzenis* in particular being responsible for the decline in winter moth populations, which are now maintained at very low equilibrium levels by the parasitoids. The situation in Wytham is quite different. The winter moth is often abundant but rarely defoliates the trees, and *Cyzenis*, although present, has almost no impact on the winter moth population. *Cyzenis* would thus seem to be a most unlikely candidate for a biological control agent.

This intriguing situation is fortunately one into which we can delve more deeply using the data provided by Varley and Gradwell and Embree. Their studies point to a most significant difference between the two winter moth populations, a difference that lies in their respective pupal mortalities. In Wytham, there is a very high winter moth pupal mortality in the soil (average of 72% over 17 years) due to beetles and to some extent shrews (Frank, 1967a, b; Buckner, 1969; East, 1974; Kowalski, 1977). Furthermore, the mortality is markedly density dependent, as shown by the solid circles in Figure 9.1. The *Cyzenis* pupae in Wytham, which are in the soil for some 10 months, suffer a similar mortality, thought to be largely due to the same predators (Figure 9.1, hollow circles). This, too, is likely to show a density dependent relationship with winter moth

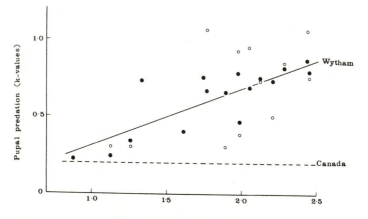

Log$_{10}$ winter moth larvae / m^2

FIGURE 9.1. The relationship between the pupal mortality of both the winter moth and *Cyzenis* ascribed to predation (expressed as *k*-values) and the density of winter moth larvae each year. (●) Winter moth mortality in Wytham (Varley and Gradwell, 1968); (O) *Cyzenis* mortality in Wytham. The broken line is the assumed winter moth and *Cyzenis* pupal mortality in Nova Scotia (Embree, 1966). The solid line is the regression for the Wytham winter moth data ($y = 0.37x$; $p < 0.01$) (from Hassell, 1977).

density, although this is not confirmed from the data in Figure 9.1. In contrast, Embree's studies point to a much lower winter moth pupal mortality in Nova Scotia (approximately 35%) without any hint of density dependence (Figure 9.1, broken line). We shall assume that *Cyzenis* in Nova Scotia is similarly affected.

It is now possible to demonstrate in a simple way the importance of this pupal mortality, and the other mortalities that have been identified, to the winter moth equilibrium level in the two countries. We commence yet again with the Nicholson-Bailey model, but now modified to include the density dependent pupal mortalities in Wytham and all the other density independent mortalities known to occur:

$$N_{t+1} = \lambda M_n [\alpha N_t^{1-b}] \exp(-aP_t)$$
$$P_{t+1} = M_p [\alpha N_t^{1-b}] [1 - \exp(-aP_t)], \tag{9.2}$$

where αN_t^{1-b} is the density dependent pupal mortality of the form shown by the regression in Figure 9.1, and M_n and M_p are the total density independent mortalities acting on the host and parasitoid respectively. The equilibrium winter moth population N^* is obtained from equation (9.2) in the usual way by setting $N_{t+1} = N_t = N^*$ and $P_{t+1} = P_t = P^*$, to give the expression

$$A = \alpha M_p a = \frac{1}{N^*} \frac{[\log_e (\alpha M_n \lambda) - b \log_e N^*]}{[(N^*)^{-b} - 1/(\alpha M_n \lambda)]}. \tag{9.3}$$

The term $\alpha M_p a$ may now be identified as the parasitoid efficiency A after correction for the density dependent pupal mortality, just as $\lambda M_n(\alpha N_t^{1-b})$ is the corrected host rate of increase allowing for all the host mortalities.

Estimates for the values of the parameters in equation (9.3) that are appropriate to Wytham and to Nova Scotia are listed in Table 9.1, and enable the relationships between N^* and A in Figure 9.2 to be drawn. The Wytham

TABLE 9.1. Values of parameters in equation (9.2) that are appropriate to Wytham (data from Varley, Gradwell, and Hassell, 1973) and Nova Scotia (data from Embree, 1965). (Asterisk denotes values for Nova Scotia that are assumed to be the same as in Wytham).

| | Value for: | |
Parameter	Wytham	Nova Scotia
b	0.37	0
λ	75	75*
αM_n	0.06	0.016
αM_p	0.38	0.63
a	0.075 m²	0.075 m²*
$A = \alpha M_p a$	0.028	0.047

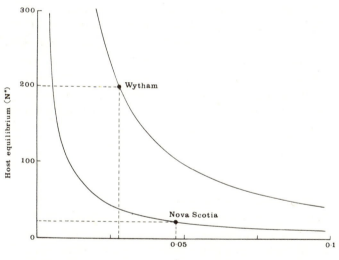

A

FIGURE 9.2. The relationships between the host equilibrium N^* and the overall parasitoid efficiency corrected for pupal mortality (A in equation (9.3)). The two curves represent slices through the surface in Figure 2.2 appropriate to the different effective rates of increase of the winter moth in Wytham and Nova Scotia. The points on each curve have been fixed using the parameter values in Table 9.1 to determine the appropriate values of A (from Hassell, 1977).

and Nova Scotia curves in the figure are different because the two average life tables are different, at least in terms of b, M_n, and M_p. We are looking, therefore, at two slices through Figure 2.2, corresponding to the different effective rates of increase of the winter moth in the two regions. As a final step, the estimates of *Cyzenis* efficiency A from Table 9.1 are used to obtain the winter moth equilibrium populations from the appropriate curve as shown by the dotted lines in the figure. The result is a winter moth equilibrium in Wytham that is ten times as great as in Nova Scotia, a difference that would be enhanced were the analysis for Nova Scotia extended to include the effects of *Agrypon* as well as *Cyzenis*.

This analysis is really a cautionary tale. It shows that an ineffective parasitoid in one locality (*Cyzenis* in Wytham) may have spectacular effects (*Cyzenis* in Nova Scotia) when liberated from restraining mortalities within the system. With such subtleties sometimes determining the outcome of biological control programs, it will be difficult to escape completely from the present ad hoc basis of natural enemy introductions.

STABILITY

Our discussion on winter moth equilibrium levels has taken no heed of whether the equilibrium is stable. In fact, while equation (9.2) applied to the Wytham data has a stable equilibrium due to the density dependent pupal predation, for the Nova Scotia data without any density dependence it behaves just like a Nicholson-Bailey model—with increasing oscillations. This is in marked contrast to the situation actually observed in Nova Scotia where the winter moth has been a rare insect since 1963 (Embree, 1971). Indeed, it is a characteristic of effective biological control projects that the host-parasitoid equilibria appear quite stable without periodic pest outbreaks. An important element in the understanding of biological control is therefore to identify the factors that may be responsible for this stability.

First, we can dismiss a density dependent rate of increase of the host population as the sole cause of stability. Recalling equation (2.12) and Figure 2.7, we can see that a Nicholson-Bailey model extended to include host density dependence permits no more than about a 40% depression of the host equilibrium consistent with local stability (i.e. $q = N^*/K \simeq 0.4$) (Beddington, Free, and Lawton, 1975). That such depressions are trifling compared to those observed from successful biological control projects has recently been stressed by Beddington, Free, and Lawton (1978). They have shown that in all cases where there is adequate data on pest population levels before and after completely successful biological control programs, the value of q does not exceed 0.025 (see Figure 9.3). We may conclude, therefore, that prior to biological control, it is likely that resource limitation of the host will be an important stabilizing factor. But after successful biological control, this can no longer be invoked, and we should now seek those properties of the parasitoid-host interaction itself which can lead to extremely low equilibrium levels being stable.

Of the responses considered in earlier chapters, three can contribute to stability: the sigmoid functional response, predator aggregation, and mutual interference. Of these, neither sigmoid responses nor mutual interference are likely to stabilize the system at very low equilibrium levels (i.e. $q \ll 1$). The sigmoid response alone cannot stabilize a coupled, discrete generation interaction (see page 45), and Beddington, Free, and Lawton (1978) go further in showing that it is unlikely to be important in obtaining very small values for q. Similarly, we have seen from Chap-

173

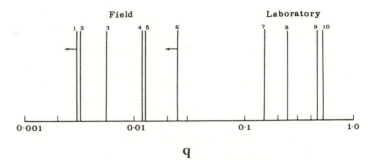

FIGURE 9.3. Some values of $q(= N*/K)$ from field (1 to 6) and laboratory (7 to 10) studies of parasitoid-host interactions. (1) *Aphytis melinus-Aonidiella aurantii;* (2) *Trioxys pallidus-Chromaphis juglandicola;* (3) *Aphytis maculicornis-Parlatoria oleae;* (4) *Aphytis melinus-Aonidiella aurantii;* (5) *Cyzenis albicans-Operophtera brumata;* (6) *Olesicampe benefactor-Pristiphora erichsonii;* (7) *Nasonia vitripennis-Musca domestica;* (8) *Neocatolaccus mamezophagus-Callosobruchus chinensis;* (9) *Nemeritis canescens-Anagasta kühniella;* (10) *Heterospilus prosopidus-Callosobruchus chinensis* (from Beddington, Free, and Lawton (1978), in which the source references are given).

ter 5 that mutual interference at equilibrium population levels is likely to have only a small effect on stability.

In contrast, the differential exploitation of host patches as a result of parasitoids or predators aggregating in regions of high host density can be of the greatest importance to stability, even at the lowest of q-values (Beddington, Free, and Lawton, 1978). This important point is illustrated most simply from the detail independent model of predator aggregation from Chapter 4 (see Figure 4.10). With random search ($k \rightarrow \infty$), we reobtain Figure 2.7 in which no stable equilibrium is possible for small values of q. This situation changes little until non-random search is such that $k \leq 1$, whereupon the model becomes stable for arbitrarily small values of q. A simulation illustrating this is shown in Figure 9.4. Thus, predator aggregation can stabilize an interaction, however small the equilibrium populations. Interestingly, Figure 9.3 also shows that the range of q-values from laboratory parasitoid-host systems fall into quite a different category from the successful biological control results, the q-values in the four available examples varying from about 0.15 to 0.5. These are a little closer to the predictions from the Nicholson-Bailey model with host density dependence (equation (2.12)), and it is probably significant that, at least in three of the four systems, the hosts were not sufficiently uneven in their distribution for non-random search to be the stabilizing influence.

In short, an ideal parasitoid for biological control, at least in perennial crop systems, should have a high search rate and be able to aggregate in patches of high host density. The high search rate leads to a greater depression of the host equilibrium and the aggregation ensures that this equilibrium will be stable. In this context it is interesting that the parasitoid *Cyzenis albicans* discussed in the previous

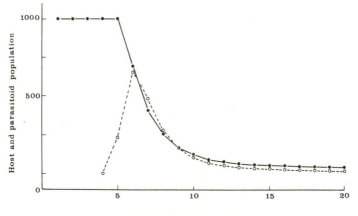

FIGURE 9.4. A numerical simulation using equation (4.19) to illustrate the powerful stabilizing effect of a parasitoid when $k \ll 1$. The host population is assumed to have a maximum size (1,000) above which it cannot rise. Assumed parameter values: $\lambda = 2$; $a = 0.75$; $k = 0.15$.

175

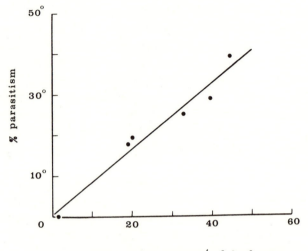

FIGURE 9.5. The relationship between the percentage parasitism (angular transformation) of winter moth larvae by *Cyzenis albicans* and the winter moth larval density on different trees (after Hassell, 1968).

section, apart from having a fairly high search rate, is known to search non-randomly and to cause appreciably greater parasitism on those trees with more plentiful winter moth larvae, as shown in Figure 9.5. It thus seems likely that herein lies the explanation for the very low and stable winter moth equilibrium in Nova Scotia over the past ten years or more.

SOME RELATIONSHIPS BETWEEN PARAMETERS

It is clear that the reproductive success of parasitoids (and to a considerable extent of predators also) depends critically on the number of hosts parasitized (or prey eaten) which itself is largely determined by such parameters as longevity T, handling time T_h, search rate a or a', interference m or bT_w, and aggregation in patches of high host density μ. It is unlikely, therefore, that the values of these various parameters for a particular parasitoid species are simply a potpourri mix bearing no relationship to each other. They should all relate in some way to the details of life history and to the searching strategy adopted. The information available is, unfortunately, still fragmentary and comes from only a handful of species. Thus, while we can speculate on plausible relationships, the data permit few firm conclusions to be made.

Longevity and Handling Time

A parasitoid has nothing to gain in having a long handling time. On the contrary, a short handling time increases the time available for search and hence the likelihood of finding further hosts. Selection should therefore act to reduce the handling time for laying a single egg as much as possible. The limits to this will be imposed mainly by the type of host parasitized. Some larval parasitoids have a handling time of less than a minute, while in species that must drill through a hard chorion or pupal covering, or drill into the host's feeding chamber, it may be well over an hour. To look at handling time alone, however, is misleading: it is the ratio T_h/T that should be minimized. Thus, an unavoidably long handling time can be compensated by a long searching life T; and indeed we have already seen signs of this from Table 3.1 where in all cases $T_h/T \ll 1$. In short, we would expect parasitoids with long handling times to be either relatively long-lived or to have gregarious larvae and hence lay many eggs in a single host.

Searching Efficiency, Interference, Aggregation, and the Degree of Polyphagy

The relationships between searching efficiency, interference, aggregation, and the range of host species accepted have been the subject of considerable speculation. Using laboratory data from a variety of parasitoid species, Rogers and Hassell (1974) showed that interference and searching efficiency, expressed as m and Q respectively from equation (5.2), were positively correlated; Rogers and Hubbard (1974) that m and the host range were negatively correlated; and Hassell (1977) that Q and the host range were also negatively related. Unfortunately, reexamination of these results with the inclusion of several further data points renders the relationships no longer significant. We are forced, therefore, one step backward and must only speculate that pronounced interference and a restricted host range are more likely to be found in species with a relatively high searching efficiency.

There are sound biological reasons to expect this. The most fundamental of these is the notion that parasitoids with a very narrow host range are more likely than polyphagous ones to respond to specific cues from a particular host species. The several examples of parasitoids making use of host pheromones or sound in locating their hosts are relevant here (see page 53). There are several likely consequences of such efficient host location.

(1) The searching efficiency a' for a particular host species will be higher than in a polyphagous parasitoid.
(2) Patches of high host density will be more efficiently discovered, leading to a higher value of the aggregation index μ.
(3) More efficient aggregation will lead to more frequent parasitoid encounters and hence perhaps to more mutual interference bT_w.

We now have a plausible framework that relates the inherent searching efficiency to the degree of aggregation in patches of high host density and perhaps also to interference, and links all of these to the degree of polyphagy.

This is directly relevant to biological control in perennial crop systems, since we have already seen that a high searching efficiency leads to reduced equilibrium populations and that pronounced aggregation is necessary for this equilibrium to be stable. In addition, although mutual interference may be negligible at equilibrium, it could well be important following a parasitoid's introduction by enhancing its dispersal. That these three desirable attributes are likely to be most marked in relatively specific parasitoid species is strong support for the continued use of such natural enemies in biological control programs.

The situation in annual crop systems is quite different. Under these regularly disrupted conditions, the concept of an equilibrium population is less appropriate. Much more important is that the natural enemies should be efficient colonizers and capable of rapid reproduction, attributes that will often be found in polyphagous species. This point has been convincingly made by Force (1972, 1974), Ehler and van den Bosch (1974), Ehler (1977), and Dowell and Horn (1977).

MULTIPLE INTRODUCTIONS

A contentious issue among biological control workers is that of single or multiple introductions. Turnbull and Chant (1961) and Turnbull (1967) in particular have taken the view that biological control is best served by only the single "best" species being introduced. This would require preliminary studies to be undertaken in order to rank the candidate species so that the best of them could be first introduced. On the other hand, van den Bosch (1968) and Huffaker (1971) believe firmly that there is little to be lost and much to be gained by introducing a sequence of natural enemies if initially only partial control is achieved. This view was supported by Hassell and Varley (1969) who used

BIOLOGICAL CONTROL

their interference model to show that a sequence of introductions should lead to successive reductions in the host equilibrium population, as shown in Figure 8.2.

It is now possible to provide a firmer theoretical basis for the practice of multiple introductions, making use of the two parasitoid-one host model of Chapter 8, namely,

$$N_{t+1} = \lambda N_t f_1(P_t) f_2(Q_t)$$
$$P_{t+1} = N_t [1 - f_1(P_t)] \qquad (9.4)$$
$$Q_{t+1} = N_t f_1(P_t) [1 - f_2(Q_t)],$$

where f_1 and f_2 are as defined in equation (8.3). The equilibrium and stability boundaries of this model are displayed in Figures 8.3 and 8.4a and b for various values of k_1 and k_2.

Let us consider the situation where a single parasitoid (species P) has been successfully introduced to combat a pest and now exists in a stable interaction. This is the situation modeled by equation (4.19) where $k < 1$, mimicking a parasitoid that tends to aggregate in patches of high host density. The host equilibrium population, however, is not sufficiently depressed and a further parasitoid (species Q) is to be introduced, with the aim of significantly lowering the host equilibrium and yet keeping the interaction stable. To remain in accord with equation (9.4), we will assume that Q either acts on the hosts surviving from P, or acts at the same time as P but is the inferior larval competitor (see page 149).

Leaving aside for the moment the extent to which the host equilibrium becomes reduced, there are several possibilities.

(1) *Q becomes established and coexists with P.* This is the desired outcome, illustrated by the simulation in Figure 9.6a, and occurs in region F of Figures 8.3 and 8.4a. It is thus most likely to be achieved if Q is strongly stabilizing

180

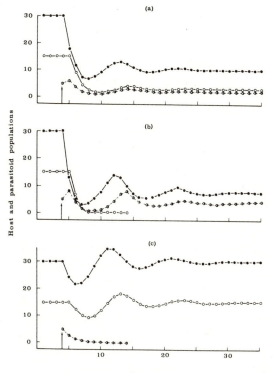

FIGURE 9.6. Numerical simulations from equation (9.4) to illustrate three of the possible outcomes following the introduction of a second parasitoid species (Q) where a stable parasitoid-host equilibrium already exists $(P^* = 15; N^* = 30)$. (a) comes from region F of Figure 8.3 where $\lambda = 2$; $a_1 = 0.25$; $a_2 = 0.35$; $k_1 = k_2 = 0.25$. (b) comes from region A of Figure 8.3 where $\lambda = 2$; $a_1 = 0.1$; $a_2 = 0.4$; $k_1 = k_2 = 0.5$. (c) comes from region B of Figure 8.3 where $\lambda = 2$; $a_1 = 0.1$; $a_2 = 0.05$; $k_1 = k_2 = 0.5$. Each of these is further discussed in the text. (Symbols: (●) N, (○)P, (◑)Q).

$(k_2 \ll 1)$ and if Q has a higher searching efficiency than $P(a_2 > a_1)$, since the conditions for local stability are now at their most extensive, as shown in Figure 8.3. This outcome is well seen from several biological control programs, a notable example being the control of the olive scale, *Parlatoria oleae,* by two aphelinid parasitoids, *Aphytis maculicornis (P)* and *Coccophagoides utilis (Q)* (Huffaker and Kennett, 1966).

(2) *Q replaces P and persists in a stable interaction* (Figure 9.6b). This occurs in region *A* of Figure 8.3 and hence is only possible if *Q* is stabilizing ($k_2 < 1$) and has a considerably higher searching efficiency than *P*. This outcome will also be beneficial since the higher searching efficiency of *Q* will lead to some reduction in the host equilibrium. An example which appears to illustrate this is shown in Figure 9.7 and comes from the introduction of several species of braconid parasitoids in the genus *Opius* for the control of the fruit-fly, *Dacus dorsalis.*

(3) *Q fails to become established, leaving P in its former stable interaction* (Figure 9.6c). This occurs in region *B* of Figures 8.3 and 8.4a and hence is only likely if *Q* has a smaller searching efficiency than *P*, irrespective of whether it searches randomly ($k_2 \rightarrow \infty$) or in a stabilizing way ($k_2 < 1$).

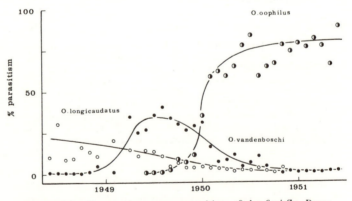

FIGURE 9.7. Changes in percent parasitism of the fruitfly, *Dacus dorsalis,* by three species of *Opius* parasitoids. Note that each successive parasitoid species appears to cause a higher level of parasitism (data from Bess, van den Bosch, and Haramoto (1961), after Varley, Gradwell, and Hassell, 1973).

It is a frequent outcome in biological control, but the only loss is one of wasted effort.

(4) *Q forces P to extinction and then remains in an unstable single parasitoid-host interaction*. This occurs in region *C* of Figure 8.3 and in *A* and *C* of Figure 8.4a and so is only feasible if *Q* is destabilizing ($k > 1$) and has a higher searching efficiency than *P*.

(5) *Both Q and P can persist only in an unstable three-species interaction*. This occurs in region *E* of Figures 8.3 and 8.4a and is thus best avoided by *Q* being strongly stabilizing ($k \ll 1$).

Only categories (4) and (5) are highly undesirable outcomes for a biological control project, since the practical result will be the extinction of both parasitoid species or perhaps their coexistence in a locally unstable interaction. The recipe for avoiding this is to select as the second species only parasitoids which promote stability by tending to aggregate where hosts are abundant (i.e. $\mu \gg 0$ or $k \ll 1$). Furthermore, if this property is coupled with a high searching efficiency, only the desirable categories (1) and (2) are likely. The fact that the extinction of both species as a result of the introduction of the second has not been recorded from biological control programs is encouraging, and suggests that the particular parameter combinations in (4) and (5) above are uncommon in the real world.

In short, we find that the same requirements of a high searching efficiency and aggregation in patches of high host density that are needed for very small, stable equilibrium populations in a single parasitoid-host interaction also emerge as a necessary combination for establishing a two parasitoid-host equilibrium. Furthermore, this combination is also optimal for achieving the greatest relative de-

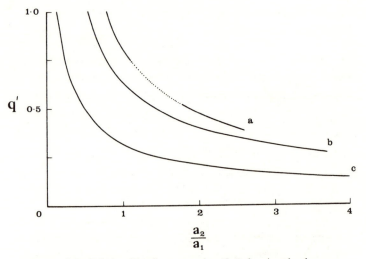

FIGURE 9.8. Relationships from equation (9.4) showing the depression in the host equilibria (q') caused by the addition of a second parasitoid (species Q) in relation to the relative searching efficiencies of $Q(=a_2)$ and $P(=a_1)$. Each curve shows a greater depression as the efficiency of Q is increased. (a) $k_1 = k_2 = 0.6$; (b) $k_1 = k_2 = 0.4$; (c) $k_1 = k_2 = 0.2$. Note that the greatest depressions are achieved with the smallest values of k, but this is counteracted by the absolute values of the host equilibrium tending to be higher as k increases. The dotted segment in curve a corresponds to a region (E in Figure 8.3) where the three-species equilibrium is always locally unstable.

pression of the host equilibrium. While some depression is unavoidable when moving from a two-species to three-species equilibrium, the greatest relative reduction will occur if, once more, parasitoid Q has the higher searching efficiency ($a_2 > a_1$) and is also strongly stabilizing ($k_2 \ll 1$). This is apparent from Figure 9.8 where q' is the ratio of the host equilibrium populations after and before the establishment of Q.

Finally, we should recall from Chapter 8 that the stable regions in Figures 8.3 and 8.4 only indicate equilibria that are *locally* stable. The initial ratio of parasitoids and hosts could thus be crucial and biological control programs fail merely because the size of the released populations of parasitoids was unsuitable.

HYPERPARASITOIDS

Biological control workers are understandably careful to prevent the accidental importation of hyperparasitoids. This caution is well-founded. Let us again assume that we have a stable single parasitoid-host interaction and that a hyperparasitoid is now released, making equation (8.9) appropriate. From Figure 8.7 we see that there is a large region where a locally stable equilibrium can occur; only if $k > 1$ or $a_2 < a_1$ may the interaction be unstable. Indeed, comparison of Figure 8.7 with Figures 8.3 and 8.4 shows that the conditions for establishing a hyperparasitoid are much broader than for a second parasitoid species. Once

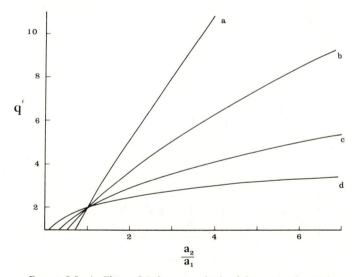

FIGURE 9.9. As Figure 9.8, but now obtained from equation (8.9) where species Q is a hyperparasitoid. The introduction of a hyperparasitoid always increases the host equilibrium. (a) $k_1 = k_2 = 0.5$; (b) $k_1 = k_2 = 0.4$; (c) $k_1 = k_2 = 0.3$; (d) $k_1 = k_2 = 0.2$.

established, the effect can only be to raise the host equilibrium as shown in Figure 9.9, where q' is now the ratio of host equilibria after and before the hyperparasitoid's introduction.

SUMMARY

Several models from earlier chapters are useful in contributing to a theoretical basis for some current biological control practices. Other things being equal, an ideal parasitoid for biological control in a perennial crop system emerges as one with a high search rate and a marked ability to aggregate in patches of high host density. The high search rate promotes low equilibrium host populations and the aggregation is necessary for this equilibrium to be stable. Such characteristics, it is argued, are likely to be most pronounced among relatively specific rather than polyphagous species.

When other things are *not* equal, predicting the outcome of a biological control venture becomes much more difficult. This is emphasized by the example of the winter moth where, in Nova Scotia, the relatively small pupal mortality is probably crucial to the effectiveness of the imported parasitoid *Cyzenis albicans*.

Interestingly, the same requirements of high searching efficiency and pronounced parasitoid aggregation also emerge as an ideal combination when one is seeking additional parasitoid species to complement those already present. The additional species will either coexist with the first species or replace it. In both cases the host equilibrium will fall and remain stable.

Models also point to the dangers of hyperparasitoids. These can be relatively easy to establish and will always have the undesired effect of raising the host equilibrium.

Epilogue

Arthropod predator-prey systems stand apart in the wealth of data they have yielded on the basic components of predation. As a result, we have good general descriptions of functional responses, aggregation, and interference, of preference and switching, and of the factors affecting predator reproduction. One by one, these components have been included in the general difference model described in Chapter 1. By adopting this stepwise procedure, the models have remained analytically tractable and the effects of each component easily discernible. The alternative, of building an ever more complicated model to which each new component is added to the last, would soon lead to an unmanageable edifice. Fortunately, the combined effects of several responses are, for the most part, those expected from the known effects of each on its own. For example, a model combining a type II functional response from equation (3.7), non-random search from equation (4.6), and mutual interference from equation (5.10), namely,

$$f(N_t, P_t) = \sum_{i=1}^{n} \left[\alpha_i \exp \left(- \frac{a'TN_t\beta_iP_t}{1 + a'T_hN_t + bT_w(\beta_iP_t - 1)} \right) \right],$$

has stability properties that depend in a straightforward way on the balance between the known stabilizing effects of the non-random search and interference, and the destabilizing effects of the functional response (Hassell and May, 1973).

Despite the plentiful information available, much more remains to be understood, both from simple single predator-single prey systems and, even more so, from more complex predator-prey communities. For instance, links should be forged between what is known of the popu-

187

lation dynamics of predators and prey in a patchy environment and the burgeoning literature on optimal foraging theory. In particular, it should be possible to determine the effects on stability of a population of predators whose aggregative responses to the density of prey per patch are described, not in the simple terms of the models of Chapter 4, but in the much more realistic vein of optimal foraging models.

There is also a need for more detailed studies on nonrandom search and switching by polyphagous predators in a patchy, multi-prey system. Such multi-species systems have been relatively neglected because of their complexity, but the advent of some very simple models that are effective in describing the outcome of non-random search opens new possibilities. They will make the modeling of multi-prey interactions, or systems with competing predators, much more manageable.

In addition to this general elaboration of the structure of interactions, true predator-prey models need to recover from their neglect at the hands of what are essentially parasitoid-host models. Parasitoids have been, and will remain, ideal laboratory subjects, but there are several questions for which they are inappropriate. The variable predator developmental rate as a function of prey density, identified in Chapter 6, could, for instance, have important dynamical consequences and lead to a departure from the standard difference equation format adopted throughout this book. The inclusion of predator and prey age structure is a further important step to take, and one likely to reveal interesting and unexpected properties.

Ideally, all such developments will continue to be characterized by a judicious blend of theory with the results from appropriate experiments.

Appendix I

Much of Chapters 2 to 5 of this book is concerned with the identification of different forms for the function $f(N_t, P_t)$ that defines the proportion of prey surviving predation in the equation

$$N_a = N_t[1 - f(N_t, P_t)], \qquad (A1.1)$$

where N_a is the total number of prey attacked by P_t predators. While several forms of (A1.1) appear frequently in the literature (e.g. Royama, 1971a; Rogers, 1971; Beddington, 1975; Hassell, 1976), their origin is less well known. This appendix, therefore, outlines the derivation of several predator attack equations which are based either on different kinds of functional response (linear, convex, or sigmoid) or on predator mutual interference. But first, however, some general recipes are outlined, which will enable all the derivations to be cast in the same mold.

USE OF THE POISSON DISTRIBUTION

One approach for obtaining equations such as (2.3), (2.4), (3.5), and (3.6) is to make use of the Poisson distribution for the occurrence of discrete, random events; in this case, the occurrence of encounters between a predator and its prey. The distribution is defined by the mean frequency of occurrence, namely, the average number of encounters with a given prey (i.e. N_e/N_t, where N_e is the total number of encounters with N_t prey). If we are dealing with parasitoids that lay an egg at each encounter, then N_e/N_t is the average number of eggs laid per host. On the other hand, for predators that consume their prey, N_e/N_t can be thought of as the average encounters with a particular "prey location" (see page 30).

With the mean defined, the probability of a host or "prey location" being encountered 0, 1, . . . n times is given by successive terms in the series

$$e^{-\bar{x}}, [e^{-\bar{x}}]\bar{x}, \ . \ . \ . \ e^{-\bar{x}} \left[\frac{\bar{x}^n}{n!}\right], \tag{A1.2}$$

where $\bar{x} = N_e/N_t$. It is, however, only the probability of a host or prey remaining undiscovered (the zero term of the distribution) that we require. Thus, the probability of actually being attacked is simply

$$1 - \exp(-N_e/N_t). \tag{A1.3}$$

Hence the total number of hosts or prey attacked N_a, becomes

$$N_a = N_t [1 - \exp(-N_e/N_t)], \tag{A1.4}$$

which is the general equation first obtained by Thompson (1924).

The final step is merely to substitute a suitable expression for N_e/N_t into (A1.4). Thus, to obtain the Nicholson-Bailey equation (2.4), we make use of equation (2.1), namely,

$$N_e/N_t = aP_t. \tag{A1.5}$$

Similarly, for the parasitoid equation (3.5), we use

$$\frac{N_e}{N_t} = \frac{a'TP_t}{1 + a'T_hN_t}, \tag{A1.6}$$

where a' is the search rate (the average encounters per prey per unit of searching time), T_h is the handling time, and T is the duration of the interaction.

A MORE GENERAL APPROACH

Alternatively, one can adopt a different and more general approach to achieve exactly the same end.

We first define $\pi(t)$ as the instantaneous encounter rate with prey; that is to say, the probability of a given prey being encountered in time dt is $\pi(t)dt$. From this, the average encounters up to time T is given by

$$\frac{N_e}{N_t} = \int_0^T \pi(t)dt, \tag{A1.7}$$

and in the special case that $\pi(t)$ remains constant during the interaction interval T, then

$$\pi(t) = \frac{N_e}{NT}. \tag{A1.8}$$

We now identify $p(t)$ as the probability of a prey *not* being attacked up to time t, namely,

$$p(t) = \frac{N_t - N_a}{N_t}, \tag{A1.9}$$

so that

$$N_a = N_t [1 - p(t)]. \tag{A1.10}$$

This paves the way for a basic equation for relating $p(t)$ and $\pi(t)$:

$$p(t + dt) = p(t) [1 - \pi(t)dt]. \tag{A1.11}$$

This states that the probability $p(t + dt)$ of not being attacked up to time $(t + dt)$ is the product of the probability of not being attacked up to t (namely, $p(t)$) times the probability of surviving the subsequent interval, dt. Expanding $p(t + dt)$, and neglecting terms of order dt^2 and higher, we get

$$p(t) + \frac{dp(t)}{dt} dt = p(t) - \pi(t)p(t)dt \tag{A1.12}$$

and hence

$$\frac{dp(t)}{p(t)} = -\pi(t)dt. \tag{A1.13}$$

191

When this is integrated, we have achieved an expression for $p(t)$ in terms of $\pi(t)$, namely,

$$p(t) \exp \left[- \int_0^T \pi(t)dt \right] \qquad (A1.14)$$

and hence, bearing (A1.10) in mind, also for N_a:

$$N_a = N_t \left[1 - \exp \left\{ - \int_0^T \pi(t)dt \right\} \right]. \qquad (A1.15)$$

Finally, from the equivalence in (A1.7), (A1.4) is reobtained, but now without recourse to the Poisson distribution.

SOME APPLICATIONS

In this section, we make use of the general recipe outlined above to derive a variety of different attack equations, beginning with that of Nicholson and Bailey and progressing to the more complex ones involving sigmoid functional responses or mutual interference among the predators.

These applications fall conveniently under two headings: (1) where prey density N_t remains constant throughout the attack period T and (2) where prey are removed as eaten and hence decline in numbers during T.

(1) N_t remains constant

This is the situation appropriate to parasitoids that can re-encounter hosts several times, at each of which a further period of handling time occurs. Alternatively, it would apply if the handling time were zero, as in the Nicholson-Bailey model, in which case this particular distinction between predators and parasitoids breaks down.

(1a) *The Nicholson-Bailey Model* ($T_h = 0$). If a' is the search rate, and P_t predators are searching, the overall encounter rate per prey is the constant $a'P_t$:

$$\pi(t) = a'P_t. \qquad (A1.16)$$

Substituting (A1.16) into (A1.7), we have

$$\frac{N_e}{N_t} = \int_0^T \pi(t)dt = a'TP_t. \qquad (A1.17)$$

Now, following the argument from (A1.9) to (A1.14), the probability of a prey not being attacked is given by

$$p(t) = \exp\left[-\int_0^T \pi(t)dt\right] = \exp(-a'TP_t), \qquad (A1.18)$$

which leads directly to the Nicholson-Bailey attack equation:

$$N_a = N_t[1 - \exp(-a'TP_t)]. \qquad (A1.19)$$

(1b) *Type II Functional Responses* $(T_h > 0)$. Recalling equation (3.4b), we now have the instantaneous encounter rate defined by

$$\pi(t) = \frac{a'P_t}{1 + a'T_hN_t}. \qquad (A1.20)$$

Thence, exactly as outlined earlier and as done for section (1a), we move to the probability of a prey not being attacked and on to the parasitoid equation (3.5):

$$N_a = N_t\left[1 - \exp\left(-\frac{a'TP_t}{1 + a'T_hN_t}\right)\right]. \qquad (A1.21)$$

(1c) *Type III Functional Responses.* The search rate is now itself a function of N_t as in equation (3.9), namely,

$$a' = \frac{bN_t}{1 + cN_t}, \qquad (A1.22)$$

where b and c are constants. Substituting for a' in (A1.20), gives

$$\pi(t) = \frac{bN_tP_t}{1 + cN_t + bT_hN_t^2}, \qquad (A1.23)$$

193

which, after following the basic recipe, leads to equation (3.11):

$$N_a = N_t \left[1 - \exp \left(- \frac{bTN_tP_t}{1 + cN_t + bT_hN_t^2} \right) \right]. \quad \text{(A1.24)}$$

(1d) *Mutual Interference among Predators.* In Chapter 5, we saw how mutual interference could be expressed in terms of b, the rate of "contacts" between predators, and T_w, the "time wasted" per contact. Thus, from equation (5.7), the encounter rate per prey becomes

$$\pi(t) = \frac{a'P_t}{1 + bT_w(P_t - 1)}. \quad \text{(A1.25)}$$

Again, in the manner of (A1.9) to (A1.15), we move to the appropriate attack equation, in this case (5.8):

$$N_a = N_t \left[1 - \exp \left(- \frac{a'TP_t}{1 + bT_w(P_t - 1)} \right) \right]. \quad \text{(A1.26)}$$

(2) N_t *no longer constant*

This section considers models for predators that consume their prey at each encounter. Alternatively, they would apply to any parasitoids that can *instantly* identify a parasitized host and thereby avoid any further expenditure of handling time. In either case, the number of available prey declines during the course of the interaction.

(2a) *Type II Functional Responses.* We commence by noting a difference from (A1.20), in that now

$$\pi(t) = \frac{a'P_t}{1 + a'T_hN(t)}, \quad \text{(A1.27)}$$

where $N(t)$ is the number of prey not attacked up to time t, in contrast to N_t which was the initial number of prey available. Recalling the definition of $p(t)$ in (A1.9) gives us the equivalence

$$N(t) = N_t p(t), \qquad (A1.28)$$

so that

$$\pi(t) = \frac{a'P_t}{1 + (a'T_h N_t)p(t)}. \qquad (A1.29)$$

Substituting (A1.29) into the basic equation (A1.13), now gives

$$\frac{dp(t)}{p(t)} = -\frac{a'P_t dt}{1 + (a'T_h N_t)p(t)}, \qquad (A1.30)$$

which, after rearranging, yields

$$\frac{[1 + (a'T_h N_t)p(t)]dp(t)}{p(t)} = -a'P_t dt. \qquad (A1.31)$$

On integrating, we have

$$\int_1^{p(t)} \left[\frac{1}{p(t)} + a'T_h N_t \right] dp(t) = -\int_0^T a'P_t dt; \qquad (A1.32)$$

that is,

$$\log_e p(t) = -a'P_t T + a'T_h N_t[1 - p(t)], \qquad (A1.33)$$

or

$$p(t) = \exp\left[-a'P_t \left\{ T - T_h \frac{N_t}{P_t}[1 - p(t)] \right\} \right]. \qquad (A1.34)$$

By definition (see A1.10),

$$N_a = N_t[1 - p(t)], \qquad (A1.35)$$

and hence, by combining this with (A1.34), we finally emerge with the predator equation (3.6):

$$N_a = N_t \left[1 - \exp\left\{ -a'P_t \left(T - T_h \frac{N_a}{P_t} \right) \right\} \right]. \qquad (A1.36)$$

(2b) *Type III Functional Responses.* As in Section (1c), these are responses where a' is no longer a constant, but a function of the prey density. However, there now exist two alternatives: that a' depends upon the initial prey density as in (A1.22), or that it depends on the prey still available at any time, in which case

$$a' = \frac{bN(t)}{1 + cN(t)}. \qquad (A1.37)$$

The former has no effect on the calculations in Section (2a), so that

$$N_a = N_t \left[1 - \exp \left\{ -\frac{bN_tP_t}{1 + cN_t} \left(T - \frac{T_hN_a}{P_t} \right) \right\} \right]. \qquad (A1.38)$$

For the second case, however, a more complex argument is involved. Remembering (A1.28), we can rewrite (A1.37) as

$$a' = \frac{bN_t p(t)}{1 + cN_t p(t)}, \qquad (A1.39)$$

so that

$$\pi(t) = \frac{bN_t p(t)P_t}{1 + cN_t p(t) + bT_hN_t^2 p(t)^2}. \qquad (A1.40)$$

Substituting this into (A1.13) and rearranging, we get

$$\frac{[1 + cN_t p(t) + bT_hN_t^2 p(t)^2] dp(t)}{p(t)^2} = -bN_tP_t dt. \qquad (A1.41)$$

Integration now leads to the following routine steps:

$$\int_1^{p(t)} \left[\frac{1}{p(t)^2} + \frac{cN_t}{p(t)} + bT_hN_t^2 \right] dp(t) = -bN_tP_t \int_0^T dt. \qquad (A1.42)$$

and

$$-\frac{1 - p(t)}{p(t)} + cN_t \left(\log_e p(t) \right) - bT_hN_t^2 (1 - p(t))$$
$$= -bN_tP_t T. \qquad (A1.43)$$

196

This is best rearranged as

$$cN_t \log_e p(t) = -bN_t P_t T + bT_h N_t^2 (1 - p(t)) + \left[\frac{1 - p(t)}{p(t)}\right].$$
(A1.44)

We now recall that $p(t) = 1 - (N_a/N_t)$, so that $1 - p(t) = N_a/N_t$ and $(1 - p(t))/p(t) = N_a/(N_t - N_a)$, thus enabling (A1.44), after some rearrangement, to be rewritten as

$$p(t) = \exp\left[-\frac{bN_t P_t}{cN_t}\left(T - \frac{T_h N_a}{P_t} - \frac{N_a}{bN_t P_t(N_t - N_a)}\right)\right].$$
(A1.45)

Finally, we make the usual step to the corresponding attack equation, in this case

$$N_a = N_t\left[1 - \exp\left\{-\frac{bP_t}{c}\left(T - \frac{T_h N_a}{P_t} - \frac{N_a}{bN_t P_t(N_t - N_a)}\right)\right\}\right].$$
(A1.46)

Appendix II

This appendix follows May (1973) in showing that the Lotka-Volterra predator-prey model, when framed in difference equations, becomes qualitatively very similar to the Nicholson-Bailey parasitoid-host model.

To demonstrate this, we commence with the Lotka-Volterra model in its more usual differential form:

$$\frac{dN_t}{dt} = N_t[r - bP_t]$$

$$\frac{dP_t}{dt} = P_t[-d + cN_t],$$

(A2.1)

where N_t and P_t are the host (= prey) and parasitoid (= predator) populations at time t, r is the intrinsic rate of increase of the hosts, d is the instantaneous parasitoid death rate in the absence of hosts, and b and c are interaction terms between the two populations. There are several means by which a difference model may be derived from its differential counterpart (May, Conway, Hassell, and Southwood, 1974). For instance, Royama (1971a) obtains a discrete Lotka-Volterra model by finding the analytical solution of the differential equation, and assigning the corresponding difference equation whose solution lies at discrete points along the continuous differential solution. This technique inevitably produces the same population trajectory (i.e. neutrally stable cycles) and hence negates the important effect of time lags. An alternative procedure is to follow May's (1973) recipe for obtaining a "homologous" difference model which incorporates the effects of time lags due to discrete population growth, attempting at the same time to preserve the essential biological features of the original model. A straightforward way of doing this would be to write the discrete Lotka-Volterra model in the form:

$$N_{t+1} - N_t = N_t[\alpha - bP_t]$$
$$P_{t+1} - P_t = P_t[-\beta + cN_t].$$

(A2.2)

Since, in the Nicholson-Bailey models all parasitoids are assumed to die between generations, we can let $\beta = 1$, in which case

$$N_{t+1} = (1 + \alpha)N_t \left[1 - \frac{b}{(1 + \alpha)} P_t \right]$$

(A2.3a)

$$P_{t+1} = cN_tP_t.$$

(A2.3b)

Two salient points emerge from (A2.3). First, it is clear that $(1 + \alpha) = \lambda$, where λ is the finite net rate of increase of the hosts. Second, both equations contain descriptions for the number of hosts parasitized, N_a. These are, from (A2.3a)

$$N_a = \frac{b}{(1 + \alpha)} N_tP_t,$$

(A2.4a)

and from (A2.3b) (since each host parasitized gives rise to one parasitoid in the next generation):

$$N_a = cN_tP_t.$$

(A2.4b)

By comparison with equation (2.1), it is now evident that

$$c = \frac{b}{(1 + \alpha)} \simeq a,$$

(A2.5)

where a is Nicholson's area of discovery. This approximation depends upon few hosts being encountered more than once (i.e. $N_e \simeq N_a$) which will occur when $N_e \ll N_t$.

The difference Lotka-Volterra model may now be written as

$$N_{t+1} = \lambda N_t[1 - aP_t]$$
$$P_{t+1} = N_t[aP_t].$$

(A2.6)

199

Comparison with the Nicholson-Bailey model (2.5), shows how the two expressions for parasitism differ:

$$f(N_t, P_t) = \exp(- aP_t) \text{ for Nicholson-Bailey} \quad (\text{A2.7})$$

and

$$f(N_t, P_t) = 1 - aP_t \text{ for Lotka-Volterra.} \quad (\text{A2.8})$$

Thus, the percentage parasitism from the difference Lotka-Volterra model is a linear, rather than exponential, function of parasitoid density, as shown by the slope B in Figure 2.1. This is a somewhat more naive view of parasitoid or predator search than that of Nicholson and Bailey, requiring perfect cooperation to avoid previously searched areas. However, Price (1970) has clearly demonstrated that some female parasitoids leave scent trails while searching that serve to prevent areas being re-searched either by the same female, or by others of the same, or even different species. Any means such as these to avoid completely random search will tend to make this difference Lotka-Volterra model more appropriate. In any case, both models have qualitatively the same unstable character (May, 1973) and will take very similar values provided that the majority of hosts are not parasitized. That is, as long as the parasitoids search a small fraction of the total area ($aP_t \ll 1$).

Appendix III

This appendix outlines the stability analysis of the general difference model (1.1), namely,

$$N_{t+1} = \lambda N_t f(N_t, P_t)$$
$$P_{t+1} = N_t[1 - f(N_t, P_t)]. \tag{A3.1}$$

It is taken from the Mathematical Appendix in Hassell and May (1973) in which a formal treatment is also given of the Nicholson-Bailey model (2.5), a handling time model (3.7), the predator aggregation model (4.6), the proportionate refuge model (4.10), and the mutual interference model (5.3). A detailed treatment of some other simple difference equation models is to be found in Maynard Smith (1968, Ch. 2), and a general account of the stability analysis of population models framed in difference equations and their relation to the corresponding analysis for systems of differential equations is given by May (1973, 1975a).

The possible time-independent equilibrium populations, N^* and P^*, from equation (1.1) are found simply by setting $N_{t+1} = N_t = N^*$ and $P_{t+1} = P_t = P^*$:

$$\lambda P^* = (\lambda - 1)N^* \tag{A3.2}$$
$$f(N^*, P^*) = \lambda^{-1}. \tag{A3.3}$$

These equations may in principle be solved to obtain N^* and P^*, provided that $\lambda > 1$.

In the real world, with its environmental vagaries, the equilibrium solution will be meaningful only if the system tends to return to these equilibrium populations whenever perturbed from them. We therefore need to know whether the equilibrium point N^*, P^* is a stable or unstable one.

To this end, we first write the perturbed populations as

$$N_t = N^*(1 + x_t) \tag{A3.4}$$
$$P_t = P^*(1 + y_t). \tag{A3.5}$$

The quantities x_t and y_t here measure the relatively small initial perturbations to prey and predator populations respectively. The dynamics of such perturbations are studied by Taylor-expanding the equation (A1.1) about the equilibrium point, and discarding terms of relative order x^2, xy, y^2 or higher, to get

$$x_{t+1} = (1 + \nu) x_t - \eta(\lambda - 1) y_t$$
$$(\lambda - 1) y_{t+1} = \lambda x_t - x_{t+1}, \qquad \text{(A3.6)}$$

where η and ν have been defined for convenience as

$$\eta \equiv -N^*(\partial f / \partial P)^* \qquad \text{(A3.7)}$$
$$\nu \equiv \lambda N^*(\partial f / \partial N)^*. \qquad \text{(A3.8)}$$

The partial derivatives of $f(N,P)$ with respect to P and N are both to be evaluated at the equilibrium point, N^*, P^*. Since in any biologically sensible model, the fraction of unparasitized hosts, namely $f(N,P)$ is likely to decrease as P increases and to increase as N increases, we expect η and ν to be non-negative. However, the formal treatment to follow encompasses arbitrary η and ν values.

As pointed out by Maynard Smith (1968), Bailey, Nicholson, and Williams (1962), and elsewhere, it is standard to write the solution for linear difference equations such as (A3.6) in the form

$$x_t = A_1(\theta_1)^t + A_2(\theta_2)^t$$
$$y_t = B_1(\theta_1)^t + B_2(\theta_2)^t. \qquad \text{(A3.9)}$$

Here the coefficients A and B are set by the initial perturbations in the generation at $t = 0$, and the time-dependence is contained solely in the factors θ^t. The quantities θ_1 and θ_2 are obtained in the usual way by substituting (A3.9) into (A3.6), to obtain relations of the form

$$(1 + \nu - \theta)A - \eta(\lambda - 1)B = 0$$
$$(\lambda - \theta)A - (\lambda - 1)\theta B = 0. \qquad \text{(A3.10)}$$

This pair of equations are consistent only if their deter-minant vanishes:

$$\det \begin{vmatrix} 1 + \nu - \theta & -\eta(\lambda - 1) \\ \lambda - \theta & -\theta(\lambda - 1) \end{vmatrix} = 0, \qquad (A3.11)$$

that is

$$\theta^2 - \theta(1 + \nu + \eta) + \lambda\eta = 0. \qquad (A3.12)$$

Thus, finally, the quantities θ_1 and θ_2 are given by

$$2\theta = (1 + \nu + \eta) \pm [(1 + \nu + \eta)^2 - 4\lambda\eta]^{1/2}. \qquad (A3.13)$$

From equation (A3.9) it is evident that the perturbations x_t and y_t will die away in time if and only if both θ_1 and θ_2 have modulus less than unity. That is, the overall stability criterion is

$$|\theta| < 1. \qquad (A3.14)$$

If the factor inside the square brackets in (A3.13) is posi-tive, both θ_1 and θ_2 are real numbers, and the damping is exponential in character; if this factor is negative, θ_1 and θ_2 are a conjugate pair of complex numbers, and the sta-bility character is oscillations of decreasing amplitude. Like-wise if $|\theta| > 1$, there ensues purely exponential growth or growing oscillations depending on whether this same factor is positive or negative.

It may be shown, after some algebraic manipulation, that application of the stability criterion (A3.14) to the specific form (A3.13) leads to the overall stability criterion.

$$\frac{\nu}{\lambda - 1} < \eta < \frac{1}{\lambda}. \qquad (A3.15)$$

In addition to this criterion, it is also required that $\eta > -(2 + \nu)/(\lambda + 1)$. However, for biologically reasonable $f(N,P)$ we expect $\nu \geqslant 0$, $\eta \geqslant 0$ so that this third condition is automatically fulfilled. Larger values of η outside the range (A3.15) lead to unstable oscillations, smaller values to unstable monotonic growth. Within the range (A3.15), the stability is oscillatory or monotonic depending on whether η is greater or less than a critical value η_0, which is the value for which the term in square brackets in (A3.13) vanishes, namely,

$$\eta_0 = (\sqrt{\lambda} - \sqrt{\lambda - 1 - \nu})^2. \qquad (A3.16)$$

The above constitutes a linearized stability analysis, valid in the neighborhood of the equilibrium point. For a class of analogous population models where growth is a continuous process, that is, where we have differential equations rather than difference equations such as (A3.1), it is possible to show that for a stable equilibrium point the global non-linear stability character is validly described by the neighborhood analysis (May, 1972). (Conversely, those models with no stable equilibrium point may possess a stable limit cycle.) The Poincaré-Bendixson techniques employed in the differential equation case have no immediate analogue for difference equations, and we have no corresponding rigorous proof that the conventional neighborhood stability analysis characterizes the global stability for very large perturbations. However, it is plausible that in the comparatively simple models referred to at the beginning of this appendix, the neighborhood analysis does indeed describe the global stability character. This conjecture is strengthened by the fact that extensive numerical studies for these models have invariably displayed the stability character predicted by the linearized analysis.

Bibliography

Akinlosotu, T. A. 1973. The role of *Diaeretiella rapae* (McIntosh) in the control of the cabbage aphid. Unpublished Ph.D. thesis, University of London.

Anderson, J. F., Hoy, M. A., and Weseloh, R. M. 1977. Field cage assessment of the potential for establishment of *Rogas indiscretus* against the gypsy moth. *Environ. Ent.*, *6*, 375–380.

Anderson, R. M. 1978. The regulation of host population growth by parasitic species. *Parasitology, 76,* 119–157.

Anderson, R. M., and May, R. M. 1978. Regulation and stability of host-parasite population interactions: I. Regulatory processes. *J. Anim. Ecol., 47,* 219–247.

Andrewartha, H. G., and Birch, L. C. 1954. *The Distribution and Abundance of Animals.* Chicago University Press, Chicago.

Arthur, A. P. 1966. Associative learning in *Itoplectis conquisitor* (Say) (Hymenoptera: Ichneumonidae). *Can. Ent., 98,* 213–223.

Askew, R. R. 1961. On the biology of the inhabitants of oak galls of Cynipidae (Hymenoptera) in Britain. *Trans. Soc. Br. Ent., 14,* 237–268.

Askew, R. R. 1971. *Parasitic Insects.* Heinemann, London.

Auslander, D., Oster, G., and Huffaker, C. 1974. Dynamics of interacting populations. *J. Franklin Inst., 297,* 345–376.

Bailey, V. A., Nicholson, A. J., and Williams, E. J. 1962. Interaction between hosts and parasites when some host individuals are more difficult to find than others. *J. Theor. Biol., 3,* 1–18.

Bakker, K., Bagchee, S. N., van Zwet, W. R., and Meelis, E. 1967. Host discrimination in *Pseudeucoila bochei* (Hymenoptera: Cynipidae). *Ent. Exp. Appl., 10,* 295–311.

Banks, C. J. 1957. The behavior of individual coccinellid larvae on plants. *Anim. Behav.*, *5*, 12–24.

Bänsch, R. 1966. On prey-seeking behavior of aphidophagous insects. *Proc. Prague Symposium Ecology of Aphidophagous Insects* (Ed. by I. Hodek), pp. 123–128. Academia, Prague.

Beddington, J. R. 1975. Mutual interference between parasites or predators and its effect on searching efficiency. *J. Anim. Ecol.*, *44*, 331–340.

Beddington, J. R., Free, C. A., and Lawton, J. H. 1975. Dynamic complexity in predator-prey models framed in difference equations. *Nature*, *225*, 58–60.

Beddington, J. R., Free, C. A., and Lawton, J. H. 1976. Concepts of stability and resilience in predator-prey models. *J. Anim. Ecol.*, *45*, 791–816.

Beddington, J. R., Free, C. A., and Lawton, J. H. 1978. Modelling biological control: on the characteristics of successful natural enemies. *Nature*, *273*, 513–519.

Beddington, J. R., and Hammond, P. S. 1977. On the dynamics of host-parasite-hyperparasite interactions. *J. Anim. Ecol.*, *46*, 811–821.

Beddington, J. R., Hassell, M. P., and Lawton, J. H. 1976. The components of arthropod predation. II. The predator rate of increase. *J. Anim. Ecol.*, *45*, 165–185.

Benson, J. F. 1973. Laboratory studies of insect parasite behaviour in relation to population models. Unpublished D.Phil. thesis, University of Oxford.

Bess, H. A., van den Bosch, R., and Haramoto, F. H. 1961. Fruit fly parasites and their activities in Hawaii. *Proc. Hawaii Ent. Soc.*, *17*, 367–378.

Birch, L. C. 1971. The role of environmental heterogeneity and genetical heterogeneity in determining distribution and abundance. *Proc. Adv. Study Inst. Dynamics Numbers Popul.* (Oosterbeek, 1970), 109–128.

Bradley, D. J., and May, R. M. 1978. Consequences of helminth aggregation for the dynamics of schistosomiasis. *Proc. Roy. Soc. Trop. Med. Hyg.*, *72*, (in press).

Broadhead, E., and Cheke, R. A. 1975. Host spatial pattern, parasitoid interference and the modelling of the dynamics of *Alaptus fusculus* (Hym.: Mymaridae), a parasitoid of two *Mesopsocus* species (Psocoptera). *J. Anim. Ecol.*, *44*, 767–793.

Brown, H. D. 1974. Defensive behavior of the wheat aphid, *Schizaphis graminum* (Rondani) (Hemiptera: Aphididae), against Coccinellidae. *J. Ent.* (A), *48*, 157–165.

Buckner, C. H. 1969. The common shrew *(Sorex araneus)* as a predator of the winter moth *(Operophtera brumata)* near Oxford, England. *Can. Ent.*, *101*, 370–375.

Burnett, T. 1951. Effects of temperature and host density on the rate of increase of an insect parasite. *Amer. Natur.*, *85*, 337–352.

Burnett, T. 1954. Influences of natural temperatures and controlled host densities on oviposition of an insect parasite. *Physiol. Zool.*, *27*, 239–248.

Burnett, T. 1956. Effects of natural temperatures on oviposition of various numbers of an insect parasite (Hymenoptera, Chalcididae, Tenthredinidae). *Ann. Ent. Soc. Am.*, *49*, 55–59.

Burnett, T. 1958a. Dispersal of an insect parasite over a small plot. *Can. Ent.*, *90*, 279–283.

Burnett, T. 1958b. A model of host-parasite interaction. *Proc. 10th Int. Congr. Ent.*, *2*, 679–686.

Burnett, T. 1977. Biological models of two acarine predators of the grain mite, *Acarus siro* L. *Can. J. Zool.*, *55*, 1312–1323.

Cade, W. 1975. Acoustically orienting parasitoids: fly phonotaxis to cricket song. *Science*, *190*, 1312–1313.

Callan, E. McC. 1944. A note on *Phanuropsis semiflaviventris* Girault (Hym., Scelionidae), an egg-parasite of cacao stink-bugs. *Proc. R. Ent. Soc. Lond. (A)*, *19*, 48–49.

Camors, F. B., and Payne, T. L. 1972. Response of *Heydenia unica* (Hymenoptera: Pteromalidae) to *Dendroctonus frontalis* (Coleoptera: Scolytidae) pheromones and a host-tree terpene. *Ann. Ent. Soc. Am., 65,* 31–33.

Caughley, G. 1976. Plant-herbivore systems. In *Theoretical Ecology: Principles and Applications.* (Ed. by R. M. May), pp. 94–113. Blackwell Scientific Publications, Oxford.

Chandler, A. E. F. 1969. Locomotory behaviour of first instar larvae of aphidophagous Syrphidae (Diptera) after contact with aphids. *Anim. Behav., 17,* 673–678.

Charnov, E. L. 1976. Optimal foraging, the marginal value theorem. *Theor. Pop. Biol., 9,* 129–136.

Chua, T. C. 1975. Population studies on the cabbage aphid, *Brevicoryne brassicae* (L.), and its parasites, with special reference to synchronization. Unpublished Ph.D. thesis, London University.

Clarke, L. R. 1963. The influence of predation by *Syrphus* sp. on the numbers of *Cardiaspina albitextura* (Psyllidae). *Aust. J. Zool., 11,* 470–487.

Clausen, C. P. 1962. *Entomophagous Insects.* Hafner, New York.

Cock, M. J. W. 1977. Searching behaviour of polyphagous predators. Unpublished Ph.D. thesis, University of London.

Cock, M. J. W. 1978. The assessment of preference. *J. Anim. Ecol., 47* (in press).

Cockrell, B. J. 1974. Prey size selection by invertebrate predators. Unpublished B.A. thesis, University of York.

Comins, H. N., and Hassell, M. P. 1976. Predation in multi-prey communities. *J. Theor. Biol., 62,* 93–114.

Connell, J. H. 1975. Some mechanisms producing structure in natural communities: a model and evidence from field experiments. In *Ecology and Evolution of Communities.* (Ed. by M. L. Cody and J. M. Diamond), pp. 460–490. Harvard University Press, Cambridge.

Cook, L. M. 1965. Oscillation in the simple logistic growth model. *Nature, 207,* 316.

Cook, R. M. 1977. Searching behaviour of various insect predators and parasitoids. Unpublished D.Phil. thesis, University of Oxford.

Cook, R. M., and Hubbard, S. F. 1977. Adaptive searching strategies in insect parasites. *J. Anim. Ecol., 46,* 115–125.

Crofton, H. D. 1971a. A quantitative approach to parasitism. *Parasitology, 63,* 179–193.

Crofton, H. D. 1971b. A model of host-parasite relationships. *Parasitology, 63,* 343–364.

Crombie, A. C. 1944. On intraspecific and interspecific competition in larvae of graminivorous insects. *J. Exp. Biol., 20,* 135–151.

Crombie, A. C. 1945. On competition between different species of graminivorous insects. *Proc. R. Soc. (B), 132,* 362–395.

Crombie, A. C. 1946. Further experiments on insect competition. *Proc. R. Soc. (B), 133,* 76–109.

DeBach, P. 1974. *Biological Control by Natural Enemies.* Cambridge University Press, London.

DeBach, P., Rosen, D., and Kennett, C. E. 1971. Biological control of coccids by introduced natural enemies. In *Biological Control* (Ed. by C. B. Huffaker), pp. 165–194. Plenum Press, New York.

DeBach, P., and Smith, H. S. 1941a. Are population oscillations inherent in the host-parasite relation? *Ecology, 22,* 363–369.

DeBach, P., and Smith, H. S. 1941b. The effect of host density on the rate of reproduction of entomophagous parasites. *J. Econ. Ent., 34,* 741–745.

DeBach, P., and Smith, H. S. 1947. Effects of parasite population density on rate of change of host and parasite populations. *Ecology, 28,* 290–298.

Dixon, A. F. G. 1959. An experimental study of the search-

209

ing behaviour of the predatory coccinellid beetle *Adalia decempunctata* (L.). *J. Anim. Ecol., 28,* 259–281.

Dixon, A. F. G. 1970. Factors limiting the effectiveness of the coccinellid beetle *Adalia bipunctata* (L.), as a predator of the sycamore aphid *Drepanosiphum platinoides* (Schr.). *J. Anim. Ecol., 39,* 739–751.

Dowell, R. V., and Horn, D. J. 1977. Adaptive strategies of larval parasitoids of the alfalfa weevil (Coleoptera: Curculionidae). *Can. Ent., 109,* 641–648.

Dransfield, R. 1975. The ecology of grassland and cereal aphids. Unpublished Ph.D thesis, University of London.

East, R. 1974. Predation on the soil dwelling stages of the winter moth at Wytham Wood, Berkshire. *J. Anim. Ecol., 43,* 611–626.

Ehler, L. E. 1977. Natural enemies of cabbage looper on cotton in the San Joaquin Valley. *Hilgardia, 45,* 73–106.

Ehler, L. E., and van den Bosch, R. 1974. An analysis of the natural biological control of *Trichoplusia ni* (Lepidoptera: Noctuidae) on cotton in California. *Can. Ent., 106,* 1067–1073.

Embree, D. G. 1965. The population dynamics of the winter moth in Nova Scotia 1954–1962. *Mem. Ent. Soc. Can., 46,* 1–57.

Embree, D. G. 1966. The role of introduced parasites in the control of the winter moth in Nova Scotia. *Can. Ent., 98,* 1159–1168.

Embree, D. G. 1971. The biological control of the winter moth in eastern Canada by introduced parasites. In *Biological Control* (Ed. by C. B. Huffaker), pp. 217–226. Plenum Press, New York.

Emlen, J. M. 1973. *Ecology: an Evolutionary Approach.* Addison-Wesley, Reading, Mass.

Evans, H. F. 1973. A study of the predatory habits of *Anthocoris* species (Hemiptera-Heteroptera). Unpublished D.Phil. thesis, University of Oxford.

Evans, H. F. 1976. Mutual interference between predatory anthocorids. *Ecological Entomology*, *1*, 283–286.

Fernando, M. H. J. P. 1977. Predation of the glasshouse red spider mite by *Phytoseiulus persimilis* A.-H. Unpublished Ph.D. thesis, University of London.

Flanders, S. E. 1959. Differential host relations of the sexes in parasitic Hymenoptera. *Ent. Exp. Appl.*, *3*, 125–142.

Fleschner, C. A. 1950. Studies on searching capacity of the larvae of three predators of the citrus red mite. *Hilgardia*, *20*, 233–265.

Force, D. C. 1972. r- and K-strategists in endemic host-parasitoid communities. *Bull. Ent. Soc. Am.*, *18*, 135–137.

Force, D. C. 1974. Ecology of insect host-parasitoid communities. *Science*, *184*, 624–632.

Fox, L. R. 1973. Food limitation, cannibalism and interactions among predators: effects on populations and communities of aquatic insects. Unpublished Ph.D. thesis, University of California, Santa Barbara.

Frank, J. H. 1967a. The insect predators of the pupal stage of the winter moth, *Operophtera brumata* (L.) (Lepidoptera: Hydriomenidae). *J. Anim. Ecol.*, *36*, 375–389.

Frank, J. H. 1967b. The effect of pupal predators on a population of winter moth, *Operophtera brumata* (L.) (Lepidoptera: Hydriomenidae). *J. Anim. Ecol.*, *36*, 611–621.

Fransz, H. G. 1974. *The Functional Response to Prey Density in an Acarine System*. P.U.D.O.C., Wageningen.

Free, C. A., Beddington, J. R., and Lawton, J. H. 1977. On the inadequacy of simple models of mutual interference for parasitism and predation. *J. Anim. Ecol.*, *46*, 543–554.

Gause, G. F. 1934. *The struggle for existence*. Williams and Wilkins, Baltimore. (Reprinted, 1964, by Hafner, New York).

Gibb, J. A. 1962. L. Tinbergen's hypothesis of the role of specific search images. *Ibis*, *104*, 106–111.

Gilpin, M. E., and Ayala, F. J. 1973. Global models of growth and competition. *Proc. Nat. Acad. Sci., 70,* 3590–3593.

Gilpin, M. E., and Justice, K. E. 1972. Reinterpretation of the invalidation of the principle of competitive exclusion. *Nature, 236,* 273–301.

Glen, D. M. 1973. The food requirements of *Blepharidopterus angulatus* (Heteroptera: Miridae) as a predator of the lime aphid, *Eucallipterus tiliae. Ent. Exp. Appl., 16,* 255–267.

Glen, D. M. 1975. Searching behaviour and prey-density requirements of *Blepharidopterus angulatus* (Fall.) (Heteroptera: Miridae) as a predator of the lime aphid, *Eucallipterus tiliae* (L.) and the leafhopper, *Alnetoidea alneti* (Dahlbom). *J. Anim. Ecol., 44,* 85–114.

Griffiths, K. J. 1969. Development and diapause in *Pleolophus basizonus* (Hymenoptera: Ichneumonidae). *Can. Ent., 101,* 907–914.

Griffiths, K. J., and Holling, C. S. 1969. A competition submodel for parasites and predators. *Can. Ent., 101,* 785–818.

Gurney, W. S. C., and Nisbet, R. M. 1978. Predator-prey fluctuations in patchy environments. *J. Anim. Ecol., 47,* 85–102.

Gutierrez, A. P. 1970a. Studies on host selection and host specificity of the aphid hyperparasite *Charips victrix* (Hymenoptera: Cynipidae). 3. Host suitability studies. *Ann. Ent. Soc. Am., 63,* 1485–1491.

Gutierrez, A. P. 1970b. Studies on host selection and host specificity of the aphid hyperparasite *Charips victrix* (Hymenoptera: Cynipidae). 4. The effect of age of host on host selection. *Ann. Ent. Soc. Am., 63,* 1491–1494.

Gutierrez, A. P. 1970c. Studies on host selection and host

specificity of the aphid hyperparasite *Charips victrix* (Hymenoptera: Cynipidae). 5. Host selection. *Ann. Ent. Soc. Am.*, *63*, 1495–1498.

Gutierrez, A. P. 1970d. Studies on host selection and host specificity of the aphid hyperparasite *Charips victrix* (Hymenoptera: Cynipidae). 6. Description of sensory structures and a synopsis of host selection and host specificity. *Ann. Ent. Soc. Am.*, *63*, 1705–1709.

Gutierrez, A. P., and van den Bosch, R. 1970a. Studies on host selection and host specificity of the aphid hyperparasite *Charips victrix* (Hymenoptera: Cynipidae). 1. Review of hyperparasitism and the field ecology of *Charips victrix*. *Ann. Ent. Soc. Am.*, *63*, 1345–1354.

Gutierrez, A. P., and van den Bosch, R. 1970b. Studies on host selection and host specificity of the aphid hyperparasite *Charips victrix* (Hymenoptera: Cynipidae). 2. The Bionomics of *Charips victrix*. *Ann. Ent. Soc. Am.*, *63*, 1355–1360.

Haldane, J. B. S. 1949. Disease and evolution. *Symposium sui fattori ecologici e genetici della speciazone negli animali. Ric. Sci.*, *19* (suppl.), 3–11.

Harcourt, D. G. 1971. Population dynamics of *Leptinotarsa decemlineata* (Say) in eastern Ontario. III. Major population processes. *Can. Ent.*, *103*, 1049–1061.

Hassell, M. P. 1968. The behavioural response of a tachinid fly *(Cyzenis albicans* (Fall.)) to its host, the winter moth *(Operophtera brumata* (L.)). *J. Anim. Ecol.*, *37*, 627–639.

Hassell, M. P. 1969. A population model for the interaction between *Cyzenis albicans* (Fall.) (Tachinidae) and *Operophtera brumata* (L.) (Geometridae) at Wytham, Berkshire. *J. Anim. Ecol.*, *38*, 567–576.

Hassell, M. P. 1971a. Mutual interference between searching insect parasites. *J. Anim. Ecol.*, *40*, 473–486.

Hassell, M. P. 1971b. Parasite behaviour as a factor contributing to the stability of insect host-parasite interactions. *Proc. Adv. Study Inst. Dynamics Numbers Popul.* (Oosterbeek, 1970), 366–379.

Hassell, M. P. 1975. Density dependence in single-species populations. *J. Anim. Ecol., 44,* 283–295.

Hassell, M. P. 1976. *The Dynamics of Competition and Predation.* Edward Arnold, London.

Hassell, M. P. 1977. Some practical implications of recent theoretical studies of host-parasitoid interactions. *Proc. XV Int. Congr. Ent.,* 608–616.

Hassell, M. P., and Comins, H. N. 1976. Discrete time models for two-species competition. *Theor. Pop. Biol., 9,* 202–221.

Hassell, M. P., and Comins, H. N. 1978. Sigmoid functional responses and population stability. *Theor. Pop. Biol., 12,* (in press).

Hassell, M. P., and Huffaker, C. B. 1969. Regulatory processes and population cyclicity in laboratory populations of *Anagasta kühniella* (Zeller) (Lepidoptera: Phycitidae). III. The development of population models. *Res. Popul. Ecol., 11,* 186–210.

Hassell, M. P., Lawton, J. H., and Beddington, J. R. 1976. The components of arthropod predation. I. The prey death-rate. *J. Anim. Ecol., 45,* 135–164.

Hassell, M. P., Lawton, J. H., and Beddington, J. R. 1977. Sigmoid functional responses by invertebrate predators and parasitoids. *J. Anim. Ecol., 46,* 249–262.

Hassell, M. P., Lawton, J. H., and May, R. M. 1976. Patterns of dynamical behaviour in single-species populations. *J. Anim. Ecol., 45,* 471–486.

Hassell, M. P., and May, R. M. 1973. Stability in insect host-parasite models. *J. Anim. Ecol., 42,* 693–736.

Hassell, M. P., and May, R. M. 1974. Aggregation in predators and insect parasites and its effect on stability. *J. Anim. Ecol., 43,* 567–594.

Hassell, M. P., and Moran, V. C. 1976. Equilibrium levels and biological control. *J. Ent. Soc. Sth. Afr., 39,* 357–366.

Hassell, M. P., and Rogers, D. J. 1972. Insect parasite responses in the development of population models. *J. Anim. Ecol., 41,* 661–676.

Hassell, M. P., and Varley, G. C. 1969. New inductive population model for insect parasites and its bearing on biological control. *Nature, 223,* 1133–1136.

Hastings, A. 1977. Spatial heterogeneity and the stability of predator-prey systems. *Theor. Pop. Biol., 12,* 37–48.

Haynes, D. L., and Sisojevic, P. 1966. Predatory behaviour of *Philodromus rufus* Walckenaer (Araneae: Thomisidae). *Can. Ent., 98,* 113–133.

Hidaka, T. 1958. Biological investigation on *Telenomus gifuensis* Ashmead (Hym.: Scelionidae), an egg-parasite of *Scotinophora lunda* Burmeister (Hem.: Pentatomidae) in Japan. *Acta Hymenopt., 1,* 75–93.

Hilborn, R. 1975. The effect of spatial heterogeneity on the persistence of predator-prey interactions. *Theor. Pop. Biol., 8,* 346–355.

Hodek, I. 1973. *Biology of Coccinellidae.* Academia, Prague.

Hokyo, N., and Kiritani, K. 1963. Two species of egg parasites as contemporaneous mortality in the egg population of the southern green stink bug, *Nezara viridula. Jap. J. Appl. Ent. Zool., 7,* 214–227.

Hokyo, N., and Kiritani, K. 1966. Oviposition behaviour of two egg parasites, *Asolcus mitsukurii* Ashmead and *Telenomus nakagawai* Watanabe (Hym., Prototrupoidea, Scelionidae). *Entomophaga, 11,* 191–201.

Holling, C. S. 1959a. The components of predation as revealed by a study of small mammal predation of the European pine sawfly. *Can. Ent., 91,* 293–320.

Holling, C. S. 1959b. Some characteristics of simple types of predation and parasitism. *Can. Ent., 91,* 385–398.

215

Holling, C. S. 1961. Principles of insect predation. *Ann. Rev. Ent., 6,* 163–182.

Holling, C. S. 1965. The functional response of predators to prey density and its role in mimicry and population regulation. *Mem. Ent. Soc. Can., 45,* 3–60.

Holling, C. S. 1966. The functional response of invertebrate predators to prey density. *Mem. Ent. Soc. Can., 48,* 1–86.

Holling, C. S. 1973. Resilience and stability of ecological systems. *Ann. Rev. Ecol. Syst., 4,* 1–24.

Hubbard, S. F. 1977. Studies on the natural control of *Pieris brassicae* with particular reference to parasitism by *Apanteles glomeratus.* Unpublished D.Phil. thesis, University of Oxford.

Huffaker, C. B. 1958. Experimental studies on predation: dispersion factors and predator-prey oscillations. *Hilgardia, 27,* 343–383.

Huffaker, C. B. (Ed.) 1971. *Biological Control.* Plenum Press, New York.

Huffaker, C. B., and Kennett, C. E. 1966. Studies of two parasites of olive scale, *Parlatoria oleae* (Colveé). IV. Biological control of *Parlatoria oleae* (Colveé) through the compensatory action of two introduced parasites. *Hilgardia, 37,* 283–335.

Huffaker, C. B., and Messenger, P. S. (Eds.) 1976. *Theory and Practice of Biological Control.* Academic Press, New York.

Huffaker, C. B., Messenger, P. S., and DeBach, P. 1971. The natural enemy component in natural control and the theory of biological control. In *Biological Control* (Ed. by C. B. Huffaker), pp. 16–67, Plenum Press, New York.

Huffaker, C. B., Shea, K. P., and Herman, S. G. 1963. Experimental studies on predation. Complex dispersion and levels of food in an acarine predator-prey interaction. *Hilgardia, 34,* 305–329.

Hussey, N. W., and Bravenboer, L. 1971. Control of pests in glasshouse culture by the introduction of natural enemies. In *Biological Control* (Ed. by C. B. Huffaker), pp. 195–216. Plenum Press, New York.

Ives, W. G. H. 1976. The dynamics of larch sawfly (Hymenoptera: Tenthredinidae) populations in southeastern Manitoba. *Can. Ent., 108,* 701–730.

Ivlev, V. S. 1961. *Experimental Ecology of the Feeding of Fishes.* Yale University Press, New Haven.

Jacobs, J. 1974. Quantitative measurement of food selection. A modification of the forage ratio and Ivlev's electivity index. *Oecologia, 14,* 413–417.

Johnson, D. M., Akre, B. G., and Crowley, P. H. 1975. Modelling arthropod predation: wasteful killing by damselfly naiads. *Ecology, 56,* 1081–1093.

de Jong, G. 1976. A model of competition for food. I. Frequency-dependent viabilities. *Amer. Natur., 110,* 1013–1027.

Kennedy, J. S. 1961. A turning point in the study of insect migration. *Nature, 189,* 785–791.

Kennedy, J. S. 1975. Insect dispersal. In *Insects, Science and Society* (Ed. by D. Pimentel), pp. 103–119. Academic Press, New York.

Kfir, R., Podoler, H., and Rosen, D. 1976. The area of discovery and searching strategy of a primary parasite and two hyperparasites. *Ecol. Ent., 1,* 287–295.

Kiritani, K. 1977. Systems approach for management of rice pests. *Proc. XV Int. Congr. Ent.,* 591–598.

Klomp, H. 1959. Infestations of forest insects and the role of parasites. *Proc. 15th Int. Congr. Zool.* (1958), 797–802.

Koebele, A. 1890. Report of a trip to Australia made under direction of the Entomologist to investigate the natural enemies of the Fluted Scale. *Bull. Bur. Ent. U.S. Dep. Agric.,* 21.

Kowalski, R. 1977. Further elaboration of the winter moth population models. *J. Anim. Ecol., 46,* 471–482.

Krebs, J. R. 1973. Behavioral aspects of predation. In *Perspectives in Ethology* (Ed. by P. P. G. Bateson and P. H. Klopfer), pp. 73–111.

Kuchlein, J. H. 1966. Mutual interference among the predacious mite *Typhlodromus longipilus* Nesbitt (Acari, Phytoseiidae). I. Effects of predator density on oviposition rate and migration tendency. *Meded. Rijksfac. LanbWet. Gent., 31,* 740–746.

Lack, D. 1954. *The Natural Regulation of Animal Numbers.* Oxford University Press, Oxford.

Laing, J. 1937. Host-finding by insect parasites. I. Observations on the finding of hosts by *Alysia manducator, Mormoniella vitripennis* and *Trichogramma evanescens. J. Anim. Ecol., 6,* 298–317.

Landenberger, D. E. 1968. Studies on selective feeding in the Pacific starfish *Pisaster* in Southern California. *Ecology, 49,* 1062–1075.

Latheef, M. A., Yeargan, K. V., and Pass, B. C. 1977. Effect of density on host-parasite interactions between *Hypera postica* (Coleoptera: Curculionidae) and *Bathyplectes anurus* (Hymenoptera: Ichneumonidae). *Can. Ent., 109,* 1057–1062.

Lawton, J. H., Beddington, J. R., and Bonser, R. 1974. Switching in invertebrate predators. In *Ecological Stability* (Ed. by M. B. Usher and M. H. Williamson), pp. 141–158. Chapman and Hall, London.

Lawton, J. H., Hassell, M. P., and Beddington, J. R. 1975. Prey death rates and rates of increase of arthropod predator populations. *Nature, 255,* 60–62.

Lindley, A. 1974. The local distribution and abundance of orbweb spiders. Unpublished D.Phil. thesis, University of Oxford.

Lotka, A. J. 1925. *Elements of Physical Biology.* Williams and

Wilkins, Baltimore. (Reissued as *Elements of Mathematical Biology* by Dover, 1956).

Luckinbill, L. S. 1973. Coexistence in laboratory populations of *Paramecium aurelia* and its predator *Didinium nasutum. Ecology, 54,* 1320–1327.

MacArthur, R. H. 1972. *Geographical Ecology.* Harper and Row, New York.

MacArthur, R. H. and Pianka, E. R. 1966. On optimal use of a patchy environment. *Amer. Natur., 100,* 603–609.

McMurtrie, R. E. 1975. Determinants of stability of large, randomly connected systems. *J. Theor. Biol., 50,* 1–11.

May, R. M. 1972. Limit cycles in predator-prey communities. *Science, 177,* 900–902.

May, R. M. 1973. On relationships among various types of population models. *Amer. Natur., 107,* 46–57.

May, R. M. 1974. Biological populations with non-overlapping generations: stable points, stable cycles, and chaos. *Science, 186,* 645–647.

May, R. M. 1975a. *Stability and Complexity in Model Ecosystems.* (Second edition). Princeton University Press, Princeton.

May, R. M. 1975b. Biological populations obeying difference equations: stable points, stable cycles, and chaos. *J. Theor. Biol., 49,* 511–524.

May, R. M. 1976a. Simple mathematical models with very complicated dynamics. *Nature, 261,* 459–467.

May, R. M. 1976b. Models for single populations. In *Theoretical Ecology: Principles and Applications.* (Ed. by R. M. May), pp. 4–25. Blackwell Scientific Publications, Oxford.

May, R. M. 1976c. Models for two interacting populations. In *Theoretical Ecology: Principles and Applications.* (Ed. by R. M. May), pp. 49–70. Blackwell Scientific Publications, Oxford.

May, R. M. 1977a. Togetherness among schistosomes: its effects on the dynamics of the infection. *Math. Biosci.*, *35*, 301–343.

May, R. M. 1977b. Predators that switch. *Nature, 269*, 103–104.

May, R. M. 1978. Host-parasitoid systems in patchy environments: A phenomenological model. *J. Anim. Ecol.*, *47* (in press).

May, R. M., Conway, G. R., Hassell, M. P., and Southwood, T. R. E. 1974. Time delays, density dependence, and single species oscillations. *J. Anim. Ecol., 43*, 747–770.

May, R. M., and Hassell, M. P. 1979. The dynamics of multiparasitoid-host interactions. (to be submitted).

May, R. M., and MacArthur, R. H. 1972. Niche overlap as a function of environmental variability. *Proc. Nat. Acad. Sci., 69*, 1109–1113.

May, R. M., and Oster, G. F. 1976. Bifurcations and dynamic complexity in simple ecological models. *Amer. Natur., 110*, 573–599.

Maynard Smith, J. 1968. *Mathematical Ideas in Biology.* Cambridge University Press, Cambridge.

Maynard Smith, J. 1974. *Models in Ecology.* Cambridge University Press, Cambridge.

Michelakis, S. 1973. A study of the laboratory interaction between *Coccinella septempunctata* larvae and its prey *Myzus persicae.* Unpublished M.Sc. thesis, University of London.

Miller, C. A. 1959. The interaction of the spruce budworm, *Choristoneura fumiferana* (Clem.), and the parasite *Apanteles fumiferanae* Vier. *Can. Ent., 91*, 457–477.

Miller, C. A. 1966. The black-headed budworm in Eastern Canada. *Can. Ent., 98*, 592–613.

Miller, J. C. 1977. Ecological relationships among parasites of *Spodoptera praefica. Environ. Ent., 6*, 581–585.

Mitchell, W. C., and Mau, R. F. L. 1971. Response of the

female Southern green stink bug and its parasite, *Trichopoda pennipes,* to male stink bug pheromones. *J. Econ. Ent., 64,* 856–859.

Mogi, M. 1969. Predation response of the larvae of *Harmonia axyridis* Pallas (Coccinellidae) to the different prey density. *Jap. J. Appl. Ent. Zool., 13,* 9–16.

Monro, J. 1975. Environmental variation and efficiency of biological control—*Cactoblastis* in the southern hemisphere. *Proc. Ecol. Soc. Australia, 9,* 204–212.

Moran, P. A. P. 1950. Some remarks on animal population dynamics. *Biometriks, 6,* 250–258.

Muesebeck, C. F. W. 1931. *Monodontomerus aereus* Walker, both a primary and secondary parasite of the brown-tail moth and the gypsy moth. *J. Agr. Res. 43,* 445–460.

Mukerji, M. K., and Le Roux, E. J. 1969. A quantitative study of food consumption and growth in *Podisus maculiventris* (Hemiptera: Pentatomidae). *Can. Ent., 101,* 387–403.

Münster-Swendsen, M., and Nachman, G. 1978. Asynchrony in insect host-parasite interaction and its effect on stability, studied by a simulation model. *J. Anim. Ecol., 47,* 159–171.

Murdie, G., and Hassell, M. P. 1973. Food distribution, searching success and predator-prey models. In *The Mathematical Theory of the Dynamics of Biological Populations* (Ed. by R. W. Hiorns), pp. 87–101. Academic Press, London.

Murdoch, W. W. 1969. Switching in general predators: experiments on predator specificity and stability of prey populations. *Ecol. Mon., 39,* 335–354.

Murdoch, W. W. 1971. The developmental response of predators to changes in prey density. *Ecology, 52,* 132–137.

Murdoch, W. W. 1973. The functional response of predators. *J. Appl. Ecol., 10,* 335–342.

Murdoch, W. W. 1977. Stabilizing effects of spatial heterogeneity in predator-prey systems. *Theor. Pop. Biol.*, *11*, 252–273.

Murdoch, W. W., Avery, S., and Smyth, M. E. B. 1975. Switching in predatory fish. *Ecology, 56*, 1094–1105.

Murdoch, W. W., and Oaten, A. 1975. Predation and population stability. *Adv. Ecol. Res., 9*, 2–131.

Nakamura, K. 1972. The ingestion in wolf spiders II. The expression of degree of hunger and amount of ingestion in relation to spider's hunger. *Res. Popul. Ecol., 14*, 82–96.

Nakasuji, F., Hokyo, N., and Kiritani, K. 1966. Assessment of the potential efficiency of parasitism in two competitive scelionid parasites of *Nezara viridula* L. (Hemiptera: Pentatomidae). *Appl. Ent. Zool., 1*, 113–119.

Nelmes, A. J. 1974. Evaluation of the feeding behaviour of *Prionchulus punctatus* (Cobb), a nematode predator. *J. Anim. Ecol., 43*, 553–565.

Nicholson, A. J. 1933. The balance of animal populations. *J. Anim. Ecol., 2*, 132–178.

Nicholson, A. J. 1947. Fluctuations of animal populations. *Rep. 26th Meeting Aust. N. Z. Assn. Advnt. Sci.*, Perth.

Nicholson, A. J. 1954. An outline of the dynamics of animal populations. *Aust. J. Zool., 2*, 9–65.

Nicholson, A. J., and Bailey, V. A. 1935. The balance of animal populations. *Part I. Proc. Zool. Soc. Lond.*, 1935, 551–598.

Noyes, J. S. 1974. The biology of the leek moth, *Acrolepia assectella* (Zeller). Unpublished Ph.D. thesis, University of London.

Oaten, A., and Murdoch, W. W. 1975a. Functional response and stability in predator-prey systems. *Amer. Natur., 109*, 299–318.

Oaten, A., and Murdoch, W. W. 1975b. Switching, functional response, and stability in predator-prey systems. *Amer. Natur., 109,* 299–318.

Oaten, A., and Murdoch, W. W. 1977. More on functional response and stability (reply to Levin). *Amer. Natur., 111,* 383–386.

Oster, G., and Guckenheimer, J. 1976. Bifurcation phenomena in population models. In *The Hopf Bifurcation and Its Applications* (Ed. by J. E. Marsden and M. McCracken), Applied Mathematical Sciences, Vol. 19. Springer-Verlag, New York.

Paine, R. T. 1966. Food web complexity and species diversity. *Amer. Natur., 100,* 65–75.

Paine, R. T. 1974. Intertidal community structure. *Oecologia, 15,* 93–120.

Pearl, R., and Reed, L. J. 1920. On the rate of growth of the population of the United States since 1790 and its mathematical representation. *Proc. Nat. Acad. Sci., 6,* 275–288.

Pielou, E. C. 1969. *An Introduction to Mathematical Ecology.* Wiley-Interscience, New York.

Prince, P. W. 1970. Trail odors: recognition by insects parasitic on cocoons. *Science, 170,* 546–547.

Prusźnyski, S. 1973. The influence of prey density on prey consumption and oviposition of *Phytoseiulus persimilis* Athias-Henriot (Acarina: Phytoseiidae). *SROB/WPRS Bulletin: Integrated Control in Glasshouses,* 1973–4, 41–46.

Pu, Che-Lung 1976. Biological control of insect pests in China. *Acta Entomologica Sinica, 19,* 247–252.

Richman, S. 1958. The transformation of energy by *Daphnia pulex. Ecol. Mon., 28,* 273–291.

Ricker, W. E. 1954. Stock and recruitment. *J. Fish. Res. Bd. Can., 11,* 559–623.

Rivard, I. 1962. Some effects of prey density on survival, speed of development, and fecundity of the predaceous mite *Melichares dentriticus* (Berl.) (Acarina: Aceosejidae). *Can. J. Zool., 40,* 1233–1236.

Roff, D. A. 1974. Spatial heterogeneity and the persistence of populations. *Oecologia, 15,* 245–258.

Rogers, D. J. 1972. Random search and insect population models. *J. Anim. Ecol., 41,* 369–383.

Rogers, D. J., and Hassell, M. P. 1974. General models for insect parasite and predator searching behaviour: interference. *J. Anim. Ecol., 43,* 239–253.

Rogers, D. J., and Hubbard, S. 1974. How the behaviour of parasites and predators promotes population stability. In *Ecological Stability* (Ed. by M. B. Usher and M. H. Williamson), pp. 99–119. Chapman and Hall, London.

Rosenzweig, M. L. 1971. Paradox of enrichment: destabilization of exploitation ecosystems in ecological time. *Science, 171,* 385–387.

Rothschild, Lord 1965. *A Classification of Living Animals.* Longmans, London.

Roughgarden, J., and Feldman, M. 1975. Species packing and predation pressure. *Ecology, 56,* 489–492.

Royama, T. 1970. Factors governing the hunting behaviour and selection of food by the great fit (*Parus major* L.). *J. Anim. Ecol., 39,* 619–668.

Royama, T. 1971a. A comparative study of models for predation and parasitism. *Res. Popul. Ecol.,* Suppl. 1, 1–91.

Royama, T. 1971b. Evolutionary significance of predator's response to local differences in prey density: a theoretical study. *Proc. Adv. Study Inst. Dynamics Numbers Popul.* (Oosterbeek, 1970), 344–357.

Safavi, M. 1968. Étude biologique et écologique des Hyménoptères parasites des oeufs des punaises des céréales. *Entomophaga, 13,* 381–495.

Salt, G. 1937. The sense used by *Trichogramma* to distinguish between parasitized and unparasitized hosts. *Proc. Roy. Soc. Lond. Serv. B, 122,* 57–75.

Sandness, J. N., and McMurtry, J. A. 1972. Prey consumption behaviour of *Amblyseius largoensis* in relation to hunger. *Can. Ent., 104,* 461–470.

Schwertfeger, F. 1935. Studien über den Mansenweschel einiger Forstschädhinge. *Z. Forst-u. Jagdw., 67,* 15–38.

Scott, A. 1920. Food of Port Erin mackeral in 1919. *Rep. Lancs. Sea-Fish. Labs, 28.*

Sechser, von B. 1970. Der Parasitenkomplex des kleinen Frostspanners (*Operophtera brumata* L.) (Lep., Geometridae) unter besondever Berücksichtigung der Kokon-Parasiten. II. Teil. *Z. angew, Ent., 66,* 144–160.

Smith, R. H., and Mead, R. 1974. Age structure and stability in models of prey-predator systems. *Theor. Pop. Biol., 6,* 308–322.

Solomon, M. E. 1949. The natural control of animal populations. *J. Anim. Ecol., 18,* 1–35.

Soper, R. S., Shewell, G. E., and Tyrrell, D. 1976. *Colcondamyia auditrix* nov. sp. (Diptera: Sarcophagidae), a parasite which is attracted by the mating song of its host, *Okanagana rimosa* (Homoptera: Cicadidae). *Can. Ent., 108,* 61–68.

Southwood, T. R. E. 1962. Migration of terrestrial arthropods in relation to habitat. *Biol. Rev., 37,* 171–214.

Southwood, T. R. E. 1975. The dynamics of insect populations. In *Insects, Science and Society* (Ed. by D. Pimentel), pp. 151–199. Academic Press, New York and London.

Southwood, T. R. E. 1976. *Ecological Methods* (Second edition). Chapman and Hall, London.

Southwood, T. R. E. 1977a. The stability of the trophic milieu, its influence on the evolution of behaviour and of responsiveness to trophic signals. *Colloques Internationaux du C.N.R.S., 265,* 471–493.

Southwood, T. R. E. 1977b. Habitat, the templet for ecological strategies? *J. Anim. Ecol., 46,* 337–365.

Southwood, T. R. E. 1977c. The relevance of population

dynamic theory to pest status. In *Origins of Pest, Parasite, Disease and Weed Problems* (Ed. by J. M. Cherrett and G. R. Sagar). pp. 35–54. Blackwell Scientific Publications, Oxford.

Southwood, T. R. E., and Comins, H. N. 1976. A synoptic population model. *J. Anim. Ecol.*, *405*, 949–965.

Spradbery, J. P. 1969. The biology of *Pseudorhyssa sternata* Merrill (Hym., Ichneumonidae), a cleptoparasite of Siricid woodwasps. *Bull. Ent. Res.*, *59*, 291–297.

Spradbery, J. P. 1970. Host finding by *Rhyssa persuasoria* (L.), an ichneumonid parasite of siricid woodwasps. *Anim. Behav.*, *18*, 103–114.

Steele, J. H. 1974. *The Structure of Marine Ecosystems*. Harvard University Press, Cambridge, Mass.

Sternlicht, M. 1973. Parasitic wasps attracted by the sex pheromones of their coccid hosts. *Entomophaga*, *18*, 339–343.

Stinner, R. E., and Lucas, H. L. 1976. Effects of contagious distributions of parasitoid eggs per host and of sampling vagaries on Nicholson's area of discovery. *Res. Popul. Ecol.*, *18*, 74–88.

Subba Rao, B. R., and Chacko, M. J. 1961. Studies on *Allophanurus indicus* n.sp., and egg parasite of *Bagrada cruciferarum* Kirkaldy (Hymenoptera: Scelionidae). *Beitr. Ent.*, *11*, 812–824.

Sullivan, D. L., and van den Bosch, R. 1971. Field ecology of the primary parasites and hyperparasites of the potato aphid *Macrosiphum euphorbiae* in the East San Francisco area. *Ann. Ent. Soc. Am.*, *64*, 389–394.

Takahashi, F. 1964. Reproduction curve with two equilibrium points: a consideration on the fluctuation of insect population. *Res. Popul. Ecol.*, *6*, 28–36.

Takahashi, F. (1968). Functional response to host density in a parasitic wasp, with reference to population regulation. *Res. Popul. Ecol.*, *10*, 54–68.

Thompson, D. J. 1975. Towards a predator-prey model incorporating age-structure: the effects of predator and prey size on the predation of *Daphnia magna* by *Ischnura elegans*. *J. Anim. Ecol.*, *44*, 907–916.

Thompson, W. R. 1924. La théorie mathématique de l'action des parasites entomophages et le facteur du hasard. *Annls. Fac. Sci. Marseille*, *2*, 69–89.

Tinbergen, L. 1960. The natural control of insects in pinewoods, 1: Factors influencing the intensity of predation by songbirds. *Arch. Neérl. Zool.*, *13*, 266–336.

Toth, R. S., and Chew, R. M. 1972. Development and energetics of *Notonecta undulata* during predation on *Culex tarsalis*. *Ann. Ent. Soc. Am.*, *65*, 1270–1279.

Townes, H. 1971. Ichneumonidae as biological control agents. *Proc. Tall Timbers Conf. on Ecological Animal Control by Habitat Management*, *3*, 235–248.

Turnbull, A. L. 1962. Quantitative studies of the food of *Linyphia triangularis* Clerck (Araneae: Linyphiidae). *Can. Ent.*, *94*, 1233–1249.

Turnbull, A. L. 1964. The searching for prey by a web-building spider *Achaearanea tepidariorum* (C. L. Kock). *Can. Ent.*, *96*, 568–579.

Turnbull, A. L. 1967. Population dynamics of exotic insects. *Bull. Ent. Soc. Am.*, *13*, 333–337.

Turnbull, A. L., and Chant, D. A. 1961. The practice and theory of biological control of insects in Canada. *Can. J. Zool.*, *39*, 697–753.

Ullyett, G. C. 1949a. Distribution of progeny by *Chelonus texanus* Cress. (Hymenoptera: Braconidae). *Can. Ent.*, *81*, 25–44.

Ullyett, G. C. 1949b. Distribution of progeny by *Cryptus inornatus* Pratt (Hymenoptera: Ichneumonidae). *Can. Ent.*, *81*, 285–299.

Utida, S. 1957. Cyclic fluctuations of population density intrinsic to the host parasite system. *Ecology*, *38*, 442–449.

van den Bosch, R. 1968. Comments on population dynamics of exotic insects. *Bull. Ent. Soc. Am., 14,* 112–115.

van den Bosch, R., and Messenger, P. S. 1973. *Biological Control.* Intext Press, New York.

van Lenteren, J. C., and Bakker, K. 1976. Functional responses in invertebrates. *Neth. J. Zool., 26,* 567–572.

van Valen, L. 1974. Predation and species diversity. *J. Theor. Biol., 44,* 19–21.

Vandermeer, J. H. 1973. On the regional stabilization of locally unstable predator-prey relationships. *J. Theor. Biol., 41,* 161–170.

Varley, G. C. 1947. The natural control of population balance in the knapweed gall-fly *(Urophora jaceana). J. Anim. Ecol., 16,* 139–187.

Varley, G. C., and Gradwell, G. R. 1960. Key factors in population studies. *J. Anim. Ecol., 29,* 399–401.

Varley, G. C., and Gradwell, G. R. 1963. The interpretation of insect population changes. *Proc. Ceylon Ass. Advmt. Sci., 18,* 142–156.

Varley, G. C., and Gradwell, G. R. 1968. Population models for the winter moth. *Symp. R. Ent. Soc. London., 4,* 132–142.

Varley, G. C., Gradwell, G. R., and Hassell, M. P. 1973. *Insect Population Ecology.* Blackwell Scientific Publications, Oxford.

Verhulst, P. F. 1838. Notice sur le loi que la population suit dans son accroissement. *Corresp. Math. Phys., 10,* 113–121.

Viktorov, G. A. 1968. The influence of the population density upon the sex ratio in *Trissolcus grandis* Thoms. (Hymenoptera, Scelionidae). *Zool. Zh., 47,* 1035–1039 (English summary).

Viktorov, G. A. 1971. Some general principles of insect

population density regulation. *Proc. XIII Int. Congr. Ent.* (Moscow), *1,* 573–576.

Viktorov, G. A., and Kotshetova, N. I. 1973. Significance of population density to the control of sex ratio in *Trissolcus volgensis* (Hymenoptera, Scelionidae). *Zool. Zh., 50,* 1753–1755.

Volterra, V. 1926. Variazioni e fluttuazioni del numero d'individui in specie animali conviventi. *Mem. Acad. Lincei., 2,* 31–113. (Translation in: Chapman, R. N. 1931. *Animal Ecology,* pp. 409–448. McGraw-Hill, New York).

Waage, J. K. 1977. Behavioural aspects of foraging in the parasitoid, *Nemeritis canescens* (Grav.). Unpublished Ph.D. thesis, University of London.

Watt, K. E. F. 1959. A mathematical model for the effect of densities of attacked and attacking species on the number attacked. *Can. Ent., 91,* 129–144.

Weseloh, R. M. 1969. Biology of *Cheiloneurus noxius,* with emphasis on host relationships and oviposition behaviour. *Ann. Ent. Soc. Am., 62,* 299–305.

Wiens, J. A. 1976. Population responses to patchy environments. *Ann. Rev. Ecol. Syst., 7,* 81–120.

Wilson, F. 1961. Adult reproductive behaviour in *Asolcus basalis* (Hymenoptera: Scelionidae). *Aust. J. Zool., 9,* 737–751.

Wratten, S. D. 1973. The effectiveness of the coccinellid beetle, *Adalia bipunctata* (L.), as a predator of the lime aphid, *Eucallipterus tiliae* L. *J. Anim. Ecol., 42,* 785–802.

Wylie, H. G. 1965. Some factors that reduce the reproductive rate of *Nasonia vitripennis* (Walk.) at high adult population densities. *Can. Ent., 97,* 970–977.

Zwölfer, H. 1971. The structure and effect of parasite complexes attacking phytophagous host insects. *Proc. Adv. Study Inst. Dynamics Numbers Popul.* (Oosterbeek, 1970), 405–418.

Author Index

231

Index to Genera

233

INDEX TO GENERA

Subject Index

age structure, 7, 9, 10, 12, 46, 48, 66, 188
aggregation, *see* predator
aggregation index μ, 62–64, 79, 178, 183
area of discovery a, 12, 30, 150, 199. *See also* search rate a or a'
attack rate a', *see* search rate a or a'

biological control, 8, 11, 26, 93–94, 150, 157, 163, 165–186

carrying capacity K, 19, 25, 77, 119, 133
chaos, 21–25, 119, 133–134
climate, 28
competition: between parasitoid larvae, 149–150, 155–157, 180; between parasitoid species, 147–164; coefficients, 133; contest, 19–20; curve, 14; discrete model (single species), 19–22, (multi-species), 133–145; nearest neighbor, 143; promiscuous, 143–144; scramble, 19–20, 22. *See also* interference, Lotka-Volterra models.
coexistence: of prey, 122, 136–145; two parasitoids–one host, 150–155, 179–183. *See also* equilibria, stability
crops: annual, 165–166, 179; perennial, 165, 175, 179
cycles, *see* limit cycles, oscillations

density dependence, 18, 20, 22–27, 45, 66–67, 73, 77, 130, 169–170; in prey rate of increase, 18–19, 77, 116–117; inverse, 38. *See also* logistic, resource limitation
developmental rate, *see* predator
difference models: contrasted with differential, 8, 17, 45, 67–68, 133, 198

disc equation, 32–33, 43, 86, 111, 116, 123, 124, 193
dispersal, 60–61, 94–95

egg maturation, 113
emigration, 18, 81
epidemics, 20
equilibria, 14–15, 45, 76, 84, 88, 90, 92–93, 100, 133–135, 152–155, 161, 167–172; depression of, 15, 24–25, 77–78, 118, 157, 167, 171–176, 179–185. *See also* stability
exploitation, 33–34, 97–98
extinction, 4, 18, 25, 59–61, 136–138, 183

fecundity, *see* predator
functional responses, 13, 28–49, 107, 109, 111, 115, 123–127, 132, 187, 189; age dependent, 46–48; type I, 29–31; type II, 29, 31–38, 43, 114, 132, 187, 193–195; type III, 29, 38–46, 129–133, 173, 192–197

generalists, 46, 130, 148, 167. *See also* polyphagous predators and parasitoids

handling time T_h, 31–38, 42, 47, 56–57, 86, 90, 97, 117, 125, 176–177, 190, 192
haplodiploidy, 83
"hunting by expectation," 55–56
hyperparasitoids, 11, 147, 157–164, 185–186; facultative, 158–159

immigration, 18
interference, 11, 80–105, 115, 117–118, 136, 139, 173, 177–180, 187, 189, 192, 194; constant m, 85, 87–90, 95–97, 99–104; "pseudo-interference," 96–105

235

Library of Congress Cataloging in Publication Data

Hassell, Michael Patrick.
  The dynamics of arthropod predator-prey systems.

  (Monographs in population biology; 13)
  Bibliography: p.
  Includes indexes.
    1. Arthropoda. 2. Predation (Biology). 3. Preda-
  tion (Biology)—Mathematical models. I. Title.
  II. Series.
  QL434.8.H37   595′.2′0453   78-51169
  ISBN 0-691-08208-1
  ISBN 0-691-08215-4 pbk.